ENGINEERING ASPECTS OF SOIL EROSION, DISPERSIVE CLAYS AND LOESS

Proceedings of a Symposium sponsored by the Soil Properties Committee of the Geotechnical Engineering Division of the American Society of Civil Engineers in conjunction with the ASCE Convention in Atlantic City, New Jersey

April 29, 1987

Edited by C.W. Lovell
School of Civil Engineering,
Purdue University,
West Lafayette, Indiana
and
Richard L. Wiltshire,
Bureau of Relcamation,
Denver, Colorado

Geotechnical Special Publication No. 10

Published by the
American Society of Civil Engineers
345 East 47th Street
New York, New York 10017-2398

ABSTRACT

This collection of papers deals with selected engineering aspects of soil erosion, dispersive clays and loess. Two types of erosion are considered—surface and subsurface (piping)—with respect to both dispersive and non-dispersive soils. Two papers deal with laboratory studies, while the remainder are case studies. Degrees of erosion are related to soil type, mineralogy, gradation and compacted condition. Remedial measures involve the mixing of selected chemicals, such as alum, gypsum, and lime, with the dispersive clays. Evaluations of the success of such stabilization are visual and qualitative. Four papers address the fabric of loess and its relation to the response of in situ loess. The technique of mercury intrusion porosimetry is applied to the interpretation of pore structure. Response of loess to dynamic loading is examined as are the difficulties of determining in situ unit weights. A final paper addresses the state-of-practice for the empirical proportioning of loess slopes.

Library of Congress Cataloging-in-Publication Data

Engineering aspects of soil erosion, dispersive clays, and loess.

(Geotechnical special publication; no. 10)
Includes index.
I. Soil mechanics—Congresses. 2. Soil stabilization—Congresses. 3. Clay soils—Congresses. 4. Loess—Congresses. I. Lovell, C.W. (Charles William) II. Wiltshire, Richard L. III. American Society of Civil Engineers. Geotechnical Engineering Division. Soil Properties Committee. IV. ASCE National Convention (1987: Atlantic City, N.J.) V. Series.
TA710.A1E53 1987 624.1′5136 87-1228
ISBN 0-87262-590-7

Library of Congress Catalog Card No.: 87-1228
ISBN 0-87262-590-7
Manufactured in the United States of America.

PREFACE

The prediction of susceptibility of earthen materials to erosion by subsurface or surface flow continues to challenge geotechnical engineers. This includes the special interaction with water demonstrated by dispersive clays. Unique problems are posed by the use of loessial materials, particularly in situ. Explanations of loess behavior are usually sought in terms of the soil fabric. This special publication contains papers that were presented at a one-day symposium, entitled "Engineering Aspects of Soil Erosion, Dispersive Clays and Loess", held at the National Convention of the American Society of Civil Engineers in Atlantic City, New Jersey, on April 29, 1987.

It is the current practice of the Geotechnical Engineering Division that each paper published in a special publication be reviewed for its content and quality. These special publications are intended to reinforce the programs presented at convention sessions or specialty conferences and to contain papers that are timely or controversial to some extent. Ordinarily the reviews are carried out within a three-month period. The standards of review are essentially those for the *ASCE Journal of Geotechnical Engineering,* but the exigencies of timeliness and the need to have the publication available at the convention preclude more than one cycle of editing and revision. Therefore, it should be recognized that there are some differences in purpose between contributions to the special publications and those in the Journal. In accordance with ASCE policy, all papers published in this volume are open for discussion in the *ASCE Journal of Geotechnical Engineering,* and they are eligible for ASCE awards. Reviews of papers published in this volume were conducted by the Geotechnical Engineering Division's Committee on Soil Properties in coordination with the Division Committee on Publications. The following committee members or cooperating persons from the general membership reviewed these papers:

Robert R. Carter
Eric C. Drum
Kenneth R. Demars
Edward W. Gray, Jr.
William H. Highter
Paul C. Knodel
C. William Lovell
Richard L. Wiltshire

We thank the authors and reviewers for their time and effort in making the symposium and these proceedings possible. We would also like to acknowledge the support of the School of Civil Engineering, Purdue University, and the assistance of Catherine Ralston, and Shiela Menaker, Manager, Book Production, ASCE.

CONTENTS

*This paper was not orally presented.

THE USE OF CHEMICAL ADDITIVES TO CONTROL THE EROSIVE BEHAVIOR OF DISPERSED CLAYS

CHARLES H. McELROY[1]

ABSTRACT: Dispersed clays are highly erodible because they contain a higher percentage of dissolved sodium in their pore water than do ordinary clays. The presence of the dominant sodium ions increases the thickness of the diffused double water layer surrounding the individual clay particles. This causes the repulsive forces to exceed the attractive forces so that the individual clay particles go into suspension in the presence of water. Due to their relatively small size and their individualistic particle behavior, dispersive clays are highly erosive, even in practically still water. Numerous treatments have been proposed to control or to minimize this erosive behavior when dispersed clays are used as foundations for structures or as embankments. Chemical additives have been very successful based on more than 16 years of experience by the Soil Conservation Service. Additives used include alum, gypsum, hydrated lime, and various mixtures of hydrated lime and agricultural lime. Sampling, specimen preparation, laboratory testing, design procedures, construction specifications, quality control and field performance of each additive are discussed.

INTRODUCTION

Dispersed clays have been identified in many countries throughout the world. Their detrimental effect on the performance of dams, canals, and other hydraulic structures such as the Squaw Creek Channel shown in Figure 1 has also been well documented in the literature (Bourdeaux and Imaizumi, 1977, Stapledon and Casinader, 1977, and Sherard et al, 1972). Dispersed clays are highly erosive because they contain a higher percentage of dissolved sodium cations in their pore water than do ordinary clays. The sodium increases the thickness of the diffused double water layer surrounding the individual clay particles. These particles readily go into suspension in the presence of practically still water because the repulsive forces exceed the attractive forces in the double layer.

PHYSICAL PROPERTIES OF DISPERSED CLAY

In addition to their unusual chemical properties, dispersed clays are further defined in this paper as having at least 12% of their particles, computed on a dry weight basis, finer than 0.005 mm as deter-

[1]Head, Soil Mechanics Laboratory, U.S. Dept. of Agriculture, Soil Conservation Service, Fort Worth, TX 76115

mined by ASTM Standard Method for Particle-Size Analysis of Soils, D 422-72 (1986). They also have a plasticity index greater than 4 (Ryker, 1977) as determined by the ASTM Standard Test Method for Liquid Limit, Plastic Limit, and Plasticity Index of Soils, D 4318-84 (1986). Generally, soils that meet these criteria are usually cohesive and the particles aggregate or act as a mass because of the attraction between particles. Soils not meeting these criteria generally react as single-grained particles and would not be expected to demonstrate mass or aggregate behavior. Silts and fine, nonplastic silty sands are two examples of soils that normally behave as single-grained particles.

FIG. 1.--Dispersed Clay Erosion on the Slopes of Squaw Creek Channel, Oklahoma (Copyright ASTM. Reprinted with Permission.)

PROBLEMS ASSOCIATED WITH DISPERSED CLAYS

Internal erosion failures and severe surface erosion are two major problems that have been identified when dispersed clays are used in hydraulic structures.

Dispersed clays generally have low to very low permeability rates. As a result, the velocity of water moving through their pores is not sufficient to cause movement of the soil particles, even under very high heads. However, once a crack or other opening occurs, the dispersed clay particles go into suspension and are easily carried away with the water moving through the opening.

Dispersed clays generally contain appreciable amounts of the more active clay minerals that make them more susceptible to drying cracks. Cracks in dams may also occur as a result of differential settlement or hydraulic fracturing (Sherard, 1986). Zones of poor compaction and points of stress transfer, especially near or around any discontinuity such as a conduit, may result in sufficient openings for piping to occur in dispersed clays. Earthen dams constructed of dispersed clays can fail very rapidly by internal erosion through cracks or other openings, even under very low heads (Sherard, et al, 1972). The major defensive design measure used to combat internal erosion of dispersed clays in embankment dams is a properly designed sand filter system.

The second problem associated with dispersed clays is the development of vertical or near vertical tunnels called "jugs" on the exposed surfaces of embankments and cut slopes. These jugs may develop as a result of small drying cracks, rodent holes, openings created by decaying roots, animal and vehicle tracks, or other small surface depressions that permit rainfall or surface runoff water to collect. Jugs have been encountered that have extended to a depth of 3 m with a bottom diameter of just less than 1 m. The diameter of the top opening varies, but often is less than 25 mm. The tops of these holes often collapse when the erosion inside the jug reaches a point where the support is eroded away. Typical jug hole development is shown in Figure 2. Severe rilling and gully development on the exposed slopes can also occur. If left unrepaired, jug holes, rills, and gulleys can leave an embankment vulnerable to breaching or piping failures. These surface discontinuities also present a hazard to maintenance operations.

The presence of even lush vegetation does not prevent the jugs and other surface erosion from occurring in dispersed clays. This paper presents the experience of the Soil Conservation Service (SCS) in using chemical additives to control surface erosion of dispersed clays.

TESTING FOR DISPERSED CLAYS

Currently, four tests are used by the SCS to identify dispersed clays in the laboratory. All four require samples that have been collected and maintained at their natural water content for reliable results. Samples must be sealed in suitable moisture-proof containers for shipment and storage prior to testing. Based on the experience of the author, the four tests are listed in order of their reliability, repeatibility, and agreement with actual field performance:

1. Pinhole Test (Sherard et al, 1976)
2. Chemical Analysis of Pore Water Extract (Sherard et al, 1972)

3. Standard Test Method for Dispersive Characteristics of Clay Soil by Double Hydrometer (ASTM D 4211-83, 1986)
4. Crumb Test (Emerson, 1967)

No single test is conclusive for all soils. For identification purposes, it is always better to run all four tests, if possible. A minimum of three of these tests is recommended unless previous experience in the area dictates otherwise. When chemical additives are considered as a design alternative for treating dispersed clays, the appropriate rate of application must be determined. The pinhole test was selected as the control test for determining the amount of chemical additive needed because of its reliability, and the capability to model in the laboratory what physically occurs in the field. Details of the pinhole test precedures have previously been published (Sherard, et al, 1976). A similar procedure has been submitted to the American Society for Testing and Materials (ASTM), Committee D 18, as a proposed standard and it is currently in the balloting process.

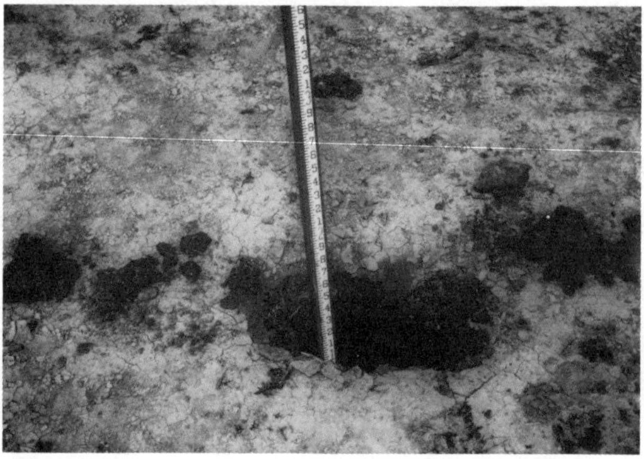

FIG. 2.--Typical Jug Hole Development (Copyright ASTM. Reprinted with Permission.)

The pinhole test is made on a specimen that has been compacted into a small permeameter using a spring-loaded tamper. The specimen is 30 to 38 mm in diameter and 38 mm long. Compacting the specimen in 5 lifts with 16 tamps per lift using a 6.8 kg spring on the Harvard Minature Compactor provides a dry unit weight approximately equal to 95% of the soil's maximum dry unit weight based on standard procedures (ASTM D 698-78, 1986). More precise control of the molding dry unit weight is not justified since the testing results are not affected by minor deviations in the dry unit weight. A 1-mm hole is punched through the specimen. Distilled water is run through this hole under heads of 50 mm, 180 mm, 380 mm, and 1020 mm. A grading system based on the turbidity of the effluent, final hole diameter, and flow rate is used to describe and categorize the results. Table 1 (Sherard, et al, 1976) summarizes these criteria. Ratings of D-1 and D-2 are considered dispersive. ND-4 and ND-3 are considered moderately dispersive. ND-2 and ND-1 are nondispersive. Dispersed clays erode rapidly and fail under the 50-mm head. Nondispersive clays do not erode even at heads of 1020 mm.

Table 1.--Summary of Criteria For Evaluating Pinhole Test Results Adopted from Sherard, et al, 1976)

Head, mm (1)	Test duration for given head, min (2)	Final flow rate ml/s (3)	Cloudiness of effluent (4)	Final hole size mm (5)	Classification (6)
50	5	1.0-1.4	Dark	≥ 2.0	D-1
50	10	1.0-1.4	Moderately dark	> 1.5	D-2
50	10	0.8-1.0	Slightly dark	≤ 1.5	ND-4
180	5	1.4-2.7	Barely visible	≥ 1.5	ND-3
380	5	1.8-3.2	Barely visible		
1020	5	>3	Clear	< 1.5	ND-2
1020	5	<3	Perfectly clear	1.0	ND-1

CHEMICAL ADDITIVES

Since the late 1960's, the SCS has used various chemical additives to improve the engineering behavior of dispersed clays. Most of our experience has been with hydrated lime (calcium hydroxide), gypsum (calcium sulfate), alum (aluminum sulfate), and mixtures of hydrated lime and agricultural lime (calcium carbonate and/or magnesium carbonate).

The amount of chemical additive needed is determined using a modification of the laboratory pinhole test. This modification consists of mixing the additive with the soil and distilled water and then allowing a curing period before molding the test specimens.

HYDRATED LIME

Hydrated lime has been mixed with dispersed clays to repair exposed surfaces of many badly eroded dams constructed since the early 1970's. It has also been used in the construction of several new dams as part of the routine design for sites containing dispersed clays. The amount of hydrated lime required is determined in the laboratory based on a series of pinhole tests. Only fresh hydrated lime should be used in preparing the specimens. Even if the lime has been stored in an air tight container, hydrated lime that is more than a month old should not be used, as it readily absorbs carbon dioxide from the atmosphere. Testing results using hydrated lime containing more than about 7% carbon dioxide can be highly erratic, and the conclusions reached may not be valid.

Usually, three or more test specimens are needed to determine the minimum amount of hydrated lime required to change the behavior of a dispersed clay (D-1, D-2, ND-3, and ND-4) to nondispersed (ND-1 or ND-2) as determined by the pinhole test. The soil sample is processed through a U.S. Standard No. 10 sieve. The material larger than the No. 10 sieve is discarded and the water content of the finer fraction is determined using standard procedures (ASTM D 2216-80, 1986). This water content is used only to determine the equivalent dry weight of the soil needed to mold the specimens. The equivalent of 200 g of dry soil for each specimen is weighed. For example, 230 g of moist soil would be required for a soil having a water content of 15%.

The amount of hydrated lime required for specimens containing 1, 2, and 3% hydrated lime by dry weight is calculated and weighed for mixing. Experience has shown that the easiest molding occurs at a water content near the ASTM D 698-78 (1986) optimum. If the natural water content of the soil is too low, distilled water is added to raise the water content to near optimum. If the natural water content is too high, air drying is necessary before molding. The addition of hydrated lime to the moist clay results in considerable drying of the mixture. Distilled water may have to be added during mixing to maintain a friable soil at a water content near optimum. After thoroughly mixing the hydrated lime and soil, the mixture is placed in an airtight container and allowed to cure for 48 hours. Additional water and further mixing may have to be employed during the curing period.

After curing, the specimens are molded into a pinhole test permeameter. The pinhole test is then run using normal procedures and grading criteria. The test specimen with the lowest percentage of hydrated lime that produces a nondispersive behavior of ND-1 or ND-2 is determined. However, because better mixing can be obtained in the laboratory than in the field, an additional 1% lime is usually added in the construction specifications to the percentage obtained from the pinhole test. SCS experience has previously been documented (McElroy, 1982) based on hundreds of tests that 2% hydrated lime will result in nondispersive behavior for nearly any dispersed clay in the pinhole test, and that 1% hydrated lime will be adequate for many clays.

Table 2 shows the results of laboratory index and pinhole tests on a typical sample of the dispersed clay used to construct Pelucia Creek Dam No. 1 located in Carroll County, Mississippi. The sample was collected from the side of a large jug hole on the severely eroded upstream slope shown in Figure 3. The test results indicated that 1% hydrated lime would change the dispersive rating of D-1 to a non-dispersive rating of ND-1 in the pinhole test. Two % hydrated lime was specified in the construction contract for repair of the dam.

TABLE 2.--Typical Laboratory Test Results From Dispersed Clay Used in the Original Construction of Pelucia Creek Dam No. 1, Carroll County, Mississippi

Sample Description (1)	Percent Passing .005 mm (2)	Percent Passing .074 mm (3)	LL (4)	PI (5)	Unified Class (6)	Pinhole Class (7)	pH (8)
Untreated borrow	18	77	29	11	CL	D-1	7.6
Treated with 1% hydrated lime	21	80	32	13	CL	ND-1	10.9
Treated with 2% hydrated lime	25	82	31	9	CL	ND-1	11.6
Treated with 3% hydrated lime	27	84	30	9	CL	ND-1	12.1

On repair jobs such as Pelucia Creek Dam No. 1, the portion of the embankment damaged by the rills, gullies, and jugs must be excavated. In most cases, this same soil will be the material that will be lime treated. The excavated soil is hauled to a nearby accessible level area where it can be mixed and cured. Topsoil and vegetation are stripped from the mixing area and stockpiled. The excavated embankment or new borrow material, if needed, is spread to a predetermined thickness in rectangular strips. The soil is scarified using a disc or other equipment to improve its workability and loosen any rocks that need to be removed to prevent damage to the rotary mixer. At this stage of the processing, the water content of the soil and lime mixture should be about 2 to 3 percentage points drier than the ASTM D 698-A (1986) optimum. This allows the lime trucks to traverse the area and spread the lime under their own power without being towed. Caution should be used when applying hydrated lime in this manner. Winds can easily blow the lime away from the immediate area creating a safety hazard and may also violate environmental rules and regulations. Most of the modern lime trucks can be regulated to control the application rate of the lime to meet the required specifications if the thickness and weight of the spread soil are known. Several other methods (Perry, 1977) have been used for quality control checks to assure that the correct percentage of lime is applied.

FIG. 3.--Severely Eroded Upstream Slope of Pelucia Creek Dam No. 1 Constructed of Dispersed Clays

Water is added to hasten the chemical reaction. The soil, lime, and water are partially mixed by discing. This prevents loss of lime due to wind or excessive rainfall. It also allows the rotary mixer to operate without spinning its wheels on a slick lime-water surface. Mixing of the lime and soil with the rotary mixer is shown in Figure 4. Water application is continued until the optimum water content is obtained.

The surface is sealed by rolling with a pneumatic roller and then allowed to cure for at least 48 hours. Water content measurements are made during the curing period to determine if additional water must be added to maintain a water content near optimum.

FIG. 4.--Mixing Dispersed Clay Soil with Hydrated Lime and Water Using a Rotary Mixer

Prior to placement of the lime-treated soil, the surfaces of the embankment are scarified and thoroughly watered. The lime-treated soil is generally hauled to the toe of the slopes and then pushed up the slope with dozers. Placement is usually accomplished in two equal lifts. Sufficient compaction can be obtained by the tracks of the dozers traversing the surfaces of each lift a minimum of two passes. The embankment surfaces are then watered as shown in Figure 5 and the topsoil replaced. The exposed surfaces are than seeded and mulched. Figure 6 shows the vegetative stand obtained after two growing seasons.

The treated layer thickness for Pelucia Creek Dam No. 1 was 380 mm. Approximately 55,500 cubic meters of treated material were used at a cost of $9.90 per cubic meter, plus the cost of the hydrated lime delivered to the site.

One of the major problems associated with the use of hydrated lime is the substantial increase in the pH value of the treated soil, to 10.5 or higher, with most falling in the range of 11 to 12. Studies conducted and reported (McElroy, 1982) show that this pH gradually declines until it approaches the original pH of the untreated soil, usually within 3 to 5 years. Table 3 presents typical data taken from the Pelucia Creek Dam No. 1 repair. These very high initial pH values make the establishment of vegetation difficult, if not impossible. Therefore, most designs include the placement of 8 to 15 cm of topsoil directly on the lime-treated surface before seeding and mulching.

FIG. 5.--Watering Embankment Slopes After Placement of Lime-Treated Dispersed Clay (Copyright ASTM. Reprinted with Permission.)

FIG. 6.--Vegetative Stand Established on Hydrated Lime Treated Soil After Two Growing Seasons

TABLE 3.--Summary of Index, Pinhole and pH Laboratory Tests on 2% Hydrated Lime Treated Soils From Pelucia Creek Dam No. 1

Sample Description (1)	Percent Passing .005 mm (2)	.074 mm (3)	LL (4)	PI (5)	Unified Class (6)	Pinhole Class (7)	pH (8)
Untreated soil	28	85	34	12	CL	D-1	7.8
Six months after treatment	24	81	31	11	CL	ND-1	11.2
Two years after treatment	21	79	30	11	CL	ND-1	9.3
Five years after treatment	20	84	29	10	CL	ND-1	8.2
Seven years after treatment	16	81	30	9	CL	ND-1	8.0

GYPSUM

In the early 1970's a small private dam located about 9 kilometers northwest of Grenada, Mississippi, experienced severe jugging and rilling of the crest and both side slopes soon after construction was completed. The owner repaired the dam in June 1975 using commercial gypsum purchased in sacks. Unfortunately, no information is available as to the amount of gypsum used nor the construction procedures used. The SCS has continued to monitor this dam and no further evidence of dispersive erosion has been observed.

On at least two other occasions, after learning about the successful Mississippi experience using gypsum, SCS laboratory results on pinhole specimens did not show an appreciable change in dispersive behavior when gypsum was added to the test specimens. One study involved gypsum percentages as high as 12% with curing times as long as three months. The specimens were prepared identically to those using hydrated lime. However, after several D-1 pinhole failures, the curing times were increased in increments of 24 hours with little or no change in the dispersive behavior in the pinhole test.

Gypsum is only slightly soluble in water. One theory is that the distilled water moved through the pinhole so rapidly in the laboratory test that the conductivity of the eroding fluid did not change and was much less than that of the pore water in the specimen. In addition, the actual surface area of the gypsum in contact with the distilled water was very small. Granulated gypsum was used in these tests as that is the form usually available commercially. The gypsum-treated pinhole specimens eroded under the 50-mm head and were classified as D-1 or D-2. Conductivity measurements of the eroding fluid and the pore water in the specimens confirmed that the conductivity of the eroding fluid was much less than that of the pore water. Gypsum does not react chemically with the clay as rapidly as does hydrated lime.

With this in mind, a series of pinhole tests were run where the gypsum was placed atop the screens placed on top of the specimens. Distilled water was allowed to percolate through the gypsum before entering the pinhole in the test specimens. The flow rate did not change appreciably, but the conductivity of the eroding fluid was close to that of the pore water. As expected, the pinhole ratings obtained under these test conditions were ND-1 and ND-2 in all tests.

A small dam in the Crabtree Creek Watershed, Wake County, North Carolina, had experienced severe jugging and rilling within one year after construction. The dam was monitored closely for two years. The number and size of the jugs continued to increase, although at a slower rate than occurred during the first year. Several methods were considered for repair with the final two alternatives being treatment with hydrated lime and some form of treatment utilizing gypsum.

Gypsum treatment was selected due to the presence of a nearby source of gypsum and the satisfactory results of the modified laboratory pinhole tests. The surface of the embankment was prepared by excavating the jug holes and backfilling with compacted material. The finished surfaces were then plated with about 7 mm of fine granulated gypsum. A disc was used to bond the gypsum into the in situ embankment soils. Topsoil was added and the vegetation planted.

After three years of close monitoring, there is no further evidence of any jugs, rills, or gullies or other surface erosion.

Additional laboratory studies are in progress at the SCS's Soil Mechanics Laboratory, Fort Worth, Texas, to learn more about the use of gypsum to treat dispersed clays.

ALUM

A common commercial grade of alluminum sulfate, alum, is used in many water treatment plants as a flocculating agent. It is highly soluble in water and readily available. For these reasons, alum appeared to be a good possibility for use in treating dispersed clays. However, alum is an acid salt (pH of 3.4 for a 1% solution) and its effect on some soils with respect to the establishment of vegetation was suspect.

The SCS decided to construct a test plot using alum as part of the Squaw Creek Channel Project repair located near Lawton, Oklahoma.

Based on the results of many pinhole, double hydrometer, and crumb laboratory tests, application rates of 0.8 and 1.6% alum applied as a 25% solution were used. Alum percentages tested ranged from 0.2% to 2% (Ryker, 1977).

A 155 meter segment of channel was treated with one-half the length being treated with 0.8% alum and the other half with 1.6%. The alum was mixed with the soil in place on the right side of the channel for a depth of 230 mm whereas the soil on the left side was stripped off for a depth of 305 mm, placed on the shoulder of the bank, treated with alum, and recompacted in two 150 mm lifts. The channel side slopes were approximately 2.5 horizontal to 1 vertical.

The 25% solution of alum and water was mixed on the site. The contractor chose to use a Hydro-seeder with a 5.7 cubic meter tank containing a large agitator near the bottom of the tank. The alum powder was furnished in bags. No problems were encountered in getting the alum into the solution. Approximately 935 cubic meters of alum treated soil were involved requiring about 18,200 kg of alum. The treated slopes were plated with 150 mm thickness of Bermuda sod-mulch and fertilized. Additional details on this test site will be discussed in a paper to be presented in this symposium (Ryker, 1987).

Recent inspections show no signs of erosive behavior of the alum-treated section of the Squaw Creek Channel.

AGRICULTURAL LIME

Lime has been used for many years to increase the fertility of soils in high rainfall areas where the soils tend to be quite acidic. This lime is often referred to as agricultural or "ag" lime and consists of ground limestone, namely calcium carbonate and/or magnesium carbonate. Agricultural lime can vary considerably depending on the type and purity of limestone used. It is generally furnished in both powder and granular forms.

Laboratory pinhole tests on specimens of dispersed clay treated with ag lime indicate that this form of lime does not reduce the erosive behavior. Ag lime contents as high as 10% have been tried. Based on the results from a limited number of tests, the powder form seems to do a better job than the granulated form, but still is not effective. To the knowledge of the author, there have been no actual field trials where ag lime alone has been successfully used to control the erosive behavior of dispersed clays.

AGRICULTURAL AND HYDRATED LIME MIXTURES

As previously discussed, about 1% extra hydrated lime is added in the construction specifications to account for waste and less efficient field mixing operations compared to controlled laboratory conditions. In remote areas, the extra cost of the additional 1% hydrated lime and the associated high transportation costs have escalated the cost of repairs. This seems to be especially true for most SCS repairs or even new projects since they do not require sufficient hydrated lime to obtain bulk prices.

A proposal was made to use agricultural lime as a carrier and mix it with a lesser amount of hydrated lime. A number of laboratory pinhole tests were run on various mixtures of ag and hydrated lime. The initial effort concerned the repair jobs for Black Creek Dams 53 and 54 in Holmes and Carroll counties, Mississippi. The results indicated that a mixture containing 20% hydrated lime and 80% ag lime based on a total lime content of 3% would be effective. Typical laboratory test results are shown in Table 4.

TABLE 4.--Typical Results of Index and Pinhole Tests on 20% Hydrated Lime and 80% Agricultural Lime (3% Total Lime) From Black Creek Dams No. 53 and 54, Mississippi

Sample Description (1)	Percent Passing 0.005 mm (2)	LL (3)	PI (4)	Unified Class (5)	Pinhole Class (6)	pH (7)
Untreated Soil	20	31	9	CL	D-1	6.9
Treated with 3% lime (20%-80% mix)	17	28	8	CL	ND-1	9.4
After six months	16	32	10	CL	ND-1	9.4
After two years	19	29	9	CL	ND-2	8.9

The lime and the dispersed clay were mixed and cured in the same manner as for hydrated lime discussed earlier. The treated soil was placed on the embankment fill, compacted, and watered. Topsoil was added and the slopes and the crest were vegetated. The thickness of the treated layer was 380 mm measured normal to the slope.

Based on these experiences and a few others, very little cost savings resulted, even though the price of the ag lime was about one-fourth that of the hydrated lime. Most of the cost of treatment is in the mixing, placement, and transportation, especially where only small quantities are involved.

Continued monitoring of these sites as well as others has shown that the ag and hydrated lime mixtures are performing well.

MONITORING OF CHEMICALLY TREATED SITES

The SCS has established a monitoring program for sites where chemical additives have been used to treat dispersed clays. Observations of field performance are made annually and samples are collected on a 3 to 5 year schedule. Complete index and laboratory dispersion tests are run on these samples. The test data, along with field observations are reported and published by the Soil Mechanics Laboratory, SCS, Fort Worth, Texas.

CONCLUSIONS

From more than 15 years of SCS experience in using various chemical additives to treat dispersed clays, the following conclusions can be drawn:

1. All sites treated with hydrated lime (calcium hydroxide) have performed well with no evidence of dispersive erosion.
2. The pinhole test is an excellent laboratory method to determine the percent of hydrated lime needed to prevent surface erosion of dispersed clays.
3. The initial pH of soils treated with hydrated lime approaches values of 10 to 12.4. However, the pH of the soils treated with hydrated lime will gradually decline to the original pH of the untreated soil within about 3 to 5 years.
4. Alum (aluminum sulfate) has also been used successfully on at least two sites to control surface erosion of dispersed clays.
5. Gypsum (calcium sulfate) has been successfully used on at least one project. Further laboratory studies and field trials are necessary before more confidence can be placed in use of gypsum to treat dispersive clays.
6. Agricultural lime (calcium carbonate and/or magnesium carbonate) is not effective as a treatment for dispersed clays, based on the pinhole test. However, when mixed with hydrated lime at a ratio of about 1:4, there is some improvement although the cost benefits over hydrated lime alone are minimal.
7. In the opinion of the author, the use of hydrated lime provides the most positive chemical treatment of dispersed clays at this time to control surface erosion.

ACKNOWLEDGEMENTS

The author wishes to express his appreciation and gratitude to the many dedicated SCS engineers, inspectors, field technicians, and laboratory testing technicians for all their assistance in making this paper possible.

APPENDIX-REFERENCES

Annual Book of ASTM Standards, (1986), Section 4, Vol. 04.08, American Society of Testing and Materials, Philadelphia, PA.

Bourdeaux, Guy, and Imaizumi, Hideo (1977), "Dispersive Clay at Sabradinho Dam", Dispersive Clays, Related Piping, and Erosion in Geotechnical Projects, STP 625, American Society for Testing and Materials, Philadelphia, PA, 13-24.

Emerson, W. W. (1977), "A Classification of Soil Aggregates Based on Their Coherence in Water", Australian Journal of Soil Research, Vol 5, Melbourne, Australia, 45-58.

McElroy, Charles H. (1982), "Effectiveness of Dispersive Clay Treatments, Progress Report, Soil Mechanics Laboratory, Soil Conservation Service, Fort Worth, TX.

Perry, Joseph P. (1977), "Lime Treatment of Dams Constructed with Dispersed Clay Soils", Transactions, American Society of Agricultural Engineers, Vol. 20, No. 6.

Ryker, N. L. (1977), "Encountering Dispersive Clays on Soil Conservation Service Projects in Oklahoma", Dispersive Clays, Related Piping, and Erosion in Geotechnical Projects, STP 625, American Society of Testing and Materials, Philadelphia, PA, 370-389.

Ryker, N. L. (1987), "Alum Treatment of Dispersive Clays in Oklahoma", Paper to be Presented at the ASCE Seminar on Engineering Aspects of Soil Erosion, Atlantic City, New Jersey.

Sherard, James L., Decker, Rey S., and Ryker, N. L. (1972), "Piping in Earth Dams of Dispersive Clay", Proceedings of the Speciality Conference on Performance of Earth and Earth - Supported Structures, Co-sponsored by the American Society of Civil Engineers and Purdue University, LaFayette, Indiana, 589-626.

Sherard, James L., Dunnigan, Lorn P., Decker, Rey S., and Steele, Edgar F. (1976), "Pinhole Test for Identifying Dispersive Soils", Journal of the Geotechnical Engineering Division, ASCE, Vol. 102, No. GT 1, 69-85.

Sherard, James L. (1986), "Hydraulic Fracturing in Embankment Dams", Journal of Geotechnical Engineering Division, ASCE, Vol. 112, No. 10, 905-927. 1986.

Stapledon, D. H., and Casinader, R. J. (1977), "Dispersive Soils at Sugarloaf Dam Site Near Melbourne, Australia", Dispersive Clays, Related Piping, and Erosion in Geotechnical Projects, STP 623, American Society for Testing and Materials, Philadelphia, PA, 432-466.

Alum Treatment of Dispersive Clays in Oklahoma

Norman L. Ryker *

Abstract

Two case histories are presented in the field trial application of alum to dispersive clay soils by the Soil Conservation Service in Oklahoma. These soil amendment trials were made on (1) the constructed banks of a channel improvement project, and on (2) the slopes of a watershed protection dam. Critically dispersive clays were identified through laboratory testing and by the presence on-site of severe erosion features common to these types of soils. At the channel project alum was applied in a solution using a hydro-seeder and then mechanically mixed with the soil to form stabilized blankets 0.23 m (9 inches) and 0.31 m (12 inches) thick. At the earth dam an alum solution was applied during concurrent construction of the dam to the outer 3.7 m (12 feet) of both slopes. In both trials alum successfully amended the erosive nature of the dispersive clays. Although application costs were less, alum is not favored over the alternate use of lime since alum is toxic to most plants and does not significantly reduce plasticity.

Introduction

Alum (aluminum sulfate, $Al_2(SO_4)_3 \times 14\ H_2O$) was used by the Soil Conservation Service (SCS) in Oklahoma for field trials in treating dispersive clay soils. The trials were made for applications on both channel banks and the surface of embankment dams. This paper describes the details of these applications and the successful results.

For many years agriculturalists have had to contend with dispersive clay soils both in normal tillage operations and in the construction of land treatment measures such as terraces, ponds, and waterways. Surface exposures of dispersive clays, sometimes called "slickspots", are produced either from the effects of sodium salt movement to the surface by capillarity, or from the removal of a topsoil covering through sheet erosion. Unless stabilized, these slickspot soils have eroded away. Organic matter, topsoil, gypsum, or limestone were sometimes applied (Harper and Plice, 1949) or, oftentimes, these areas were left idle or were returned to grass. In the early 1950's, routine laboratory testing for dispersion was being used in testing soil for constructing the larger SCS-assisted conservation practices in Oklahoma. It was not until the late 1960's that these problem soils were better identified through field testing.

* Soil Engineer, Soil Conservation Service, United States Department of Agriculture, Agricultural Center Building, Stillwater, OK 74074

Testing permitted the builder to either avoid the use of dispersive clays, to selectively place them in a noncritical zone, or to modify them using chemicals.

Hydrated lime (calcium hydroxide, $Ca(OH)_2$) has been effectively used by the SCS since 1969 (Ryker, 1976) in treating dispersive clays on watershed dams in Oklahoma. A blanket of lime-treated material provides a barrier against rainfall erosion along surface cracks. In dispersive clays such erosion typically leads to the development of piping tunnels and jugs (underground cavities which form along cracks). In this climatic area cracks are primarily caused by drying. Cracks caused by differential settlement are also of concern; however, they are usually less extensive and may require other defensive design measures in addition to lime application.

In constructing protective blankets, lime and soil are normally mixed in 0.23 m (9-inch) layers in the borrow area. After curing at high moisture content for three days, the mixture is then placed onto the scarified embankment surface in two compacted layers in an overall 0.30 - 0.38 m (12 to 15-inch) blanket thickness. Three percent lime is commonly used since, by Atterberg and shrinkage limit testing, this is usually the lowest amount that is effective in preventing drying crack formation in most Oklahoma clay soils. Curing is specified because the lime is more efficiently incorporated for two reasons: (1) unmixed lime is lost by reverting back to calcium carbonate and (2) during curing further clay aggregate breakdown occurs resulting in a better blend during final mixing and placing operations (National Lime Association, 1972).

Because of curing needs and logistical problems with timely delivery of bulk lime, few attempts have been made to build a lime-soil blanket concurrently with construction of the main earth embankment in Oklahoma. Partly for these reasons and since alum is readily soluble in water at high percentages the SCS became interested in trying this chemical as an alternate amendment for modifying clay dispersion.

Alum Testing

Alum as used herein refers to a common commercial grade of aluminum sulfate in the hydrate form, containing about 17 percent Al_2O_3. This grade is commonly used in water treatment plants, complying with the requirements of American Water Works Association standards (AWWA B403-70). Therefore, it is readily available on the commercial market. Treatment is specified in terms of percent alum which is the weight of ground hydrated aluminum sulfate (or its equivalent) in 100 units of weight of dry soil. For solutions the same relationship applies, thus a 25 percent solution is 11.35 kg (25 pounds) of alum dissolved in 45.4 kg (100 pounds) of water. On this basis, alum at 15°C (60°F) is more than 60 percent soluble in water.

In order to both treat specific problem conditions at the sites and to provide a means for field performance testing, it was decided to apply alum (in 1974) to the slopes of a flood control channel and (in 1977) to the slopes of a watershed dam. These decisions were reached following a series of tests performed at SCS soil mechanics laboratories in Lincoln, Nebraska, and in Fort Worth, Texas, supplemented with soil chemical tests performed at the SCS National Soil Survey Laboratory in Lincoln, Nebraska (Decker, 1974, Dunnigan, 1974, McElroy, 1976).

Results of pertinent test data for each of these two projects are summarized in Tables 1 and 2. Tests performed are those commonly used today for identification and classification of dispersive clay soils, namely crumb, dissolved salts, and pinhole tests (Sherard, et al., 1975, 1976), laboratory dispersion tests (ASTM, 1983) and dilution-turbidity tests (Ryker, 1976). The data in tables 1 and 2 data show that dispersive soils exist on these two structures and that these materials respond favorably to alum treatment. For the anticipated range of application to be used, alum, while lowering dispersion, does not appreciably change Atterberg and shrinkage limits and, being strongly acidic, lowers soil pH to values of 4 to 5. The effects of these low pH values on growth of desirable grass species was a principal concern in field testing. Other investigations (Bourdeaux and Imaizumi, 1976) have also shown favorable results from the addition of alum during laboratory testing of highly dispersive clays. However, the author is unaware of any field trial where alum has been used to amend dispersion or for erosion control.

Caution is advised in the storage of lime and alum for testing or for other purposes. Violent explosions are possible if these two chemicals become mixed in enclosed containers, such as in storage bins (AWWA, 1971).

Squaw Creek Channel

The Squaw Creek project is an 8 km (5 mile) long PL-566 channel improvement project located near the southeast edge of Lawton, Comanche County, Oklahoma. It is an unlined channel that, in the area selected for alum treatment, has a bottom width of 8.5 m (28 feet), a flow depth of 3.2 m (10.5 feet) and 2:1 side slopes. The channel was constructed in 1968.

During investigations made immediately prior to design and construction, laboratory tests gave a maximum of 25 percent dispersion (ASTM D-4221) from a total of 15 samples tested, although some few samples collected previously during work plan development showed the soils to be rather highly dispersed. Almost all samples collected since construction have been critically dispersive by laboratory analysis. The reasons for the low test results made prior to construction are not fully known. One possible explanation is that the test water used was demineralized but not distilled. Sampling techniques may also have been partly to blame since samples were sometimes allowed to dry prior to testing.

Soon after construction, the channel banks (Photo 1) became severely eroded by piping and jugging along several segments.

Table 1. Classification and Dispersion Tests - Squaw Creek

Sample Number or Treatment	Liquid Limit	Plasticity Index	% < 5μ	% Dispersion (1)	Crumb (2)	Pinhole (3)	T.D.S. (4)	% Sodium (5)	E.S.P. (6)	pH - Paste
72W154	30	13	26	69	4	D	23	95	31	8.1
0.2% alum	30	13		38	4	D	65	73	24	7.7
0.4% alum	30	13		30	2	I	98	65	19	7.4
0.5% alum	31	14		17	2	ND	99	66	20	7.7
3% $Ca(OH)_2$	37	10		Tr	1	ND				11.4
72W155	50	30	48	81	4	D	18	75	27	8.3
0.5% alum	51	31		26	3	D	100	64	18	7.6
0.6% alum	54	33		30	2	D	98	61	17	7.6
0.8% alum	52	31		7	2	ND	101	61	17	7.5
3% $Ca(OH)_2$	55	19		Tr	1	ND				11.6
72W156	32	15	27	33	3	D	58	79	18	8.0
0.2% alum	32	14		22	2	I	99	61	14	7.5
0.4% alum	31	14		11	1	ND	118	55	14	7.4
0.5% alum	31	13		7	1	ND	119	55	13	7.3
3% $Ca(OH)_2$	38	12		0	1	ND				11.8
75W78	53	34	40	88	4	D				
0.4% alum				53		D				
0.8% alum				13		D				
1.3% alum				8		ND				
1.6% alum				8		ND				

(1) Critical values, % >35

(2) Critical dispersion crumb values 3,4.

(3) D = dispersive, I = intermediate, ND = nondispersive

(4) Total dissolved salts, millequivalent / liter

(5) All % sodium versus T.D.S. values plot in intermediate or dispersive range (Sherard et. al., 1976)

(6) Exchangeable sodium percentage (degree of sodium saturation)

Table 2. Classification and Dispersion Test-Okfuskee Tribs Site N-9

Sample Number or Treatment	Liquid Limit	Plasticity Index	Shrinkage Limit	% < 5μ	% Dispersion	Crumb	Pinhole	pH - Paste	Dilution-Turbidity*	Chem. Test Plot
FTQ-321	56	36	10	45	58	4	D-2			
0.5% alum	59	41	7		Tr		ND			
1.0% alum	59	41	7		Tr		ND			
1.5% alum	56	39	7		Tr		ND			
2.0% alum	59	41	7		Tr		ND			
3% $Ca(OH)_2$	53	21	23		Tr		ND			
FTQ-322	38	20	13	38	13	2	ND			
0.5% alum	41	24	11		Tr		ND			
1.0% alum	41	24	11		Tr		ND			
1.5% alum	41	22	13		Tr		ND			
2.0% alum	40	23	12		Tr		ND			
F78-21	32	15		23	22	1	ND	7.2		
0.5% alum					9		ND	5.2		
0.8% alum					13		ND	4.9		
1.5% alum					4		ND	4.6		
F72-1585 (untreated borrow area sample)	33	19		31	94	3	D-1	7.4	3	D

* Critical Values : 3 and less

See Footnotes at Table 1

Numerous gullies, tunnels, cracks and jugs appeared along these lengths with an occasional piping tunnel being located as far as 12.2 m (40 feet) beyond the shoulder of the channel bank. Several side fills having pipe drop spillways were replaced following piping failures. One of these was repaired twice.

In 1972, two methods for treatment of channel banks in dispersive clays were tried on Squaw Creek for comparison and evaluation. Opposing slopes along a 152 m (500-foot) channel reach were reshaped and stabilized. A blanket of 3% hydrated lime was used on one side with the opposite slope covered with a blanket of sand-gravel

Photo 1. Typical slope condition after construction. Squaw Creek.

materials taken from a nearby pit. Both test slopes have performed excellently over the past 14 years, even though adjoining segments remained unstable, exhibiting typical erosion and jugging patterns.

As mentioned previously, alum is readily soluble and, unlike lime, a curing period is not required. Additional field trials were begun on Squaw Creek in 1975 using 0.8% and 1.6% alum. Along a 152 m (500-foot) section, one side was treated in-place 0.23 m (9 inches) deep, one-half of the length being treated at 0.8% and one-half at 1.6%. The opposing side was treated in the same proportions, 0.8% and 1.6%, however, 0.31 m (12 inches) of depth was stripped off, treated along a level strip at the top of the slope, and replaced in two 0.15 m (6-inch) increments. This work, performed through a construction contract, is described in an in-service report by Petermann, et al., (1975). All reworked surfaces, including the 1972 trials, were covered with 0.15 m (6 inches) of Bermudagrass sod mulch for surface erosion control.

Alum was delivered in 45.4 kg (100-pound) sacks. The contractor used a hydro-seeder equipped with a 5678 liter (1500 gallon) tank having an agitator and a 0.10 m (4-inch) centrifugal pump to mix alum and water to the specified 25 percent solution. Mixing was easily accomplished through vigorous action by the agitator and by recycling the pump discharge. The solution was sprayed onto surfaces prepared (and dried) by disk plowing. For the 1.6 % treatment the alum solution was applied in two increments. On the channel side where slope mixing was required, alum solution and soil were mixed using

Table 3. Most Recent of In-Place Tests (1982)

Sample Number or Treatment	% Alum	Liquid Limit	Plasticity Index	% < 5µ	% Dispersion	Crumb	Pinhole	Chem. Test	pH - Paste
Squaw Creek									
F82-1419	1.6	38	15	22	Tr	1	ND-1	ND	7.0
-1420	1.6	35	13	20	Tr	1	ND-1	ND	7.0
-1421	1.6	45	22	42	Tr	1	ND-2	D	7.1
Okfuskee Tribs. Site N-9									
F82-1416*	0.5	35	16	26	88	3	D-2	D	6.3
-1417	0.5	28	8	16	Tr	1	ND-1	ND	5.2
-1418	0.5	31	13	23	17	1	ND-1	ND	6.8

*Possible sample of topsoil layer

both a chisel plow and a 15.2 m (50-foot) drag chain equipped with 0.25 m (10-inch) spikes and a heavy steel ball on the end. As the chain was pulled parallel with the slope the ball rolled along the toe forcing the spikes into the soil which broke up the soil much like a rotary hoe.

Performance of the treated section to date has been satisfactory with no significant problems with erosion. Initially the Bermudagrass was slowed somewhat in making a complete cover due no doubt to the highly acidic condition of the underlying alum-treated blanket. There appears to be no visible difference in the areas treated, whether at the 0.8% or 1.6% level or by treatment in-place or off-site. Recent follow-up testing (Table 3) and on-site inspections (Photo 2) affirm that the tested slopes have been stabilized. Note that pH values are near neutral, most likely due to fact that the basic soils along Squaw Creek generally contain free calcium carbonate.

Okfuskee Tributaries, Site N-9

Site N-9 is an earthfill watershed (PL-566) dam located in Okfuskee County in east central Oklahoma. The dam has a maximum height of 10.7 m (35 feet), a drainage area of 1037 ha (2560 acres), and contains 91,991 m³ (120,250 cubic yards) of compacted embankment. Construction of the dam was completed in the spring of 1978.

Photo 2. Alum treated slopes. Squaw Creek.

The following observations and data affirm that the clay soils at Site N-9 were highly dispersive:

1. Erosion features typical of those in dispersive clays covered a large part of the left abutment area (Photo 3).

2. Of 19 borrow area samples collected for field dispersion testing (dilution-turbidity and crumb), prior to design, only 5 samples tested noncritical (D-T values greater than 3 and crumb test values of 1 or 2).

3. Laboratory tests on samples representing a large part of the site area were critically dispersive.

Based on the above, a decision was reached to treat the emergency spillway channel with hydrated lime and to treat the outer slopes of the embankment with alum since the use of dispersive soils could not be avoided. On the basis of laboratory tests 0.5% alum was determined to be about the lowest amount effective for this modification (Table 2). It was further decided to treat the outer 3.7 m (12 feet) of both upstream and downstream slopes with alum concurrent with construction of the interior sections. An exception was that the upstream slope below permanent pool elevation was not to be treated. All slopes were covered with a 0.15 m (6-inch) nominal thickness of untreated topsoil.

A contract for construction of Site N-9 was awarded in 1977 and the contractor began alum placement in August of that year. Like the

Photo 3. Dispersive left abutment area. Okfuskee Tribs N-9.

work on Squaw Creek, the job specifications required that alum be incorporated into the soil using a 25% solution. Unlike the Squaw Creek work, the contractor on Site N-9 was unable to mix ground alum and water at a rate that did not severely impede his other earthfill operations. Several attempts were made toward mixing the solution directly within a water wagon, however, the result was usually a substance resembling bread dough in consistency. Because of this the contractor resorted to an alternate use of liquid alum, which contained approximately 48% alum by weight.

Liquid alum was delivered to the site in tank trucks and stored in an excavated pit having a capacity of approximately 151,360 liters (40,000 gallons). This step allowed for conveniently loading of a 18,920 liter (5000 gallon) water wagon for mixing and delivery of the alum solution. A portable 7.6 cm (3-inch) pump was used to charge the water wagon as well as to mix and circulate the liquid alum and added water.

The process of actual placement began only after measurements were made to determine: (1) depth-volume relationships for the water wagon, (2) application rate, (3) travel speed, (4) soil density and (5) mixing depth (Dominick and Tyrrell, 1978). The normal procedure was to add 5676 liters (1500 gallons) of liquid alum to 10,705 liters (2829 gallons) of water and to apply this volume to 604 m (1980 feet) of lift, 0.23 m (9 inches) deep. Two coverages were made for each lift (Photos 4 and 5) with additional water added as needed to reach optimum moisture for compaction purposes (ASTM, 1978). The amount of

Photo 4. Applying alum solution

additional water for compaction was generally less than two percent. Alum distribution was double checked by use of a pH meter since the treated soil under control testing had a pH value of 4.0 to 4.5 at 0.5% alum (Figure 1).

Results and Conclusions

Although the vegetative growth on Squaw Creek was spotty until the beginning of the second growing season, no additional treatment was made except for applying fertilizer. As mentioned previously, the natural inclusion of $CaCO_3$ within the soil on Squaw Creek may have assisted in neutralizing the acidic condition caused by the alum. The alum treated blanket has performed as predicted and there has been no indication of any breakdown or reversion to a dispersed condition. On this premise, it is therefore concluded that alum, in the application rates used, is effective in treating dispersive clays like those encountered on Squaw Creek. Costs for the alum work on Squaw Creek was only slightly cheaper than the previous field tests made using hydrated lime.

Two years following alum treatment on Site N-9, a series of pH tests were made of the dam and the slickspot areas near the earth emergency spillway. Grass coverage over the dam was very weak because of the low pH values within the root depth. Some surfaces were bare but most of the dam was covered with acid-tolerant weeds such as cockleburs, jimson, and ragweeds. Based on these low pH readings, the dam was treated by broadcasting agricultural lime ($CaCO_3$) over the

Photo 5. Mixing processes after applying alum.

surface at a rate of 11.2 tonnes per hectare (5 tons per acre). The dispersive slickspots were treated with gypsum. The dam is now in good condition and well covered with Bermudagrass.

Even though at Site N-9 the costs for alum treatment, on a volume basis, were almost one-half the costs for hydrated lime treatment, there still remains a preference for lime. Lime provides for a potential reduction in plasticity and an associated reduction in shrinkage that alum does not provide. Ultimately, following the reduction in pH, there is also the apparent benefit in improved vegetative growth through the use of lime.

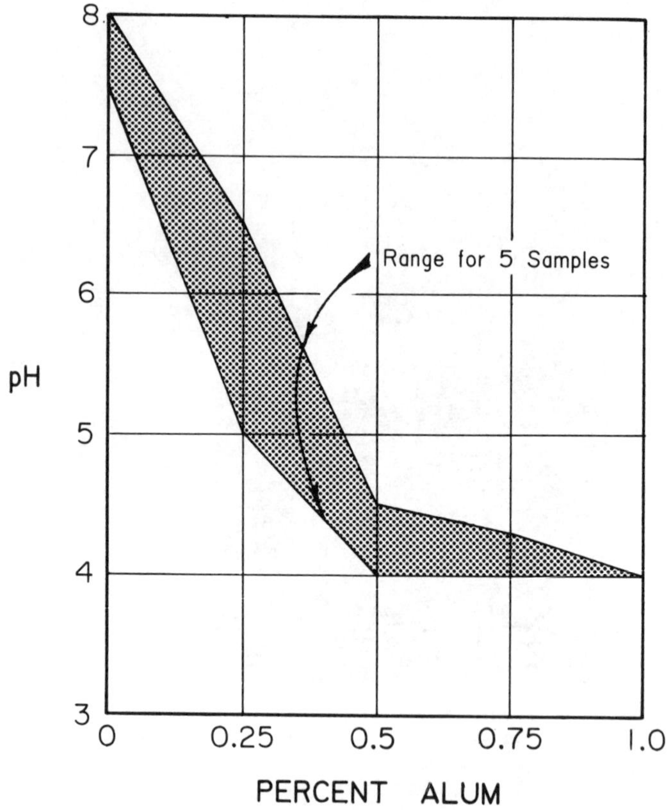

Okfuskee Tribs. N-9 - Application Control Testing

Figure 1 (Dominick et. al., 1978)

In addition to the 3.7 m (12-foot) blanket that was alum treated on the dam at Site N-9, a slickspot area downstream of the left abutment was also covered with similarly treated material. This area was not topsoiled. Table 3 shows the results of some follow-up laboratory tests made on the existing embankment. These tests are somewhat inconclusive since one or more of the samples may have come from the covering topsoil layer. Obviously, further evaluation sampling needs to be made from both the downstream area and from deeper into the dam.

Appendix

American Society for Testing and Materials, "Standard Test Method for Dispersive Characteristics of Clay Soil by Double Hydrometer", Test D 4221-83.

American Society for Testing and Materials, "Standard Test Methods for Moisture-Density Relations of Soils and Soil-Aggregate Mixtures Using 5.5-lb (2.49-kg) Rammer and 12-in. (305-mm) Drop", Test D 698-78.

American Water Works Association, "Standard for Aluminum Sulfate - Lump, Ground, or Liquid," AWWA B403-70.

American Water Works Association, "Water Quality and Treatment. A Handbook of Public Water Supplies," 3rd Ed., pg 530, McGraw-Hill Book Co., N.Y., 1971.

Bourdeaux, G. and Imaizumi, H., "Dispersive Clay at Sobradinho Dam", American Society for Testing and Materials, STP 623, 1976.

Decker, R. S., 1974, Dunnigan, L. P., 1974, and McElroy, C. H., 1976, In-Service Laboratory Reports of Test Data on Squaw Creek Channel (1974), and on Okfuskee Tributaries, Site N-9 (1976), Soil Conservation Service, USDA

Dominick, M. D. and Tyrrell, D. B., "Alum Treatment for Dispersive Soils, Site N-9, Okfuskee Tributaries", Unpublished Report, Soil Conservation Service, USDA, Stillwater, Oklahoma, January 1978.

Harper, H. J. and Plice, M. J., "Suggestions for Improving Slickspot Soils", Bulletin No. B-329, Oklahoma Agricultural Experiment Station, Oklahoma State University, February 1949.

National Lime Association, "Lime Stabilization Construction Manual", Bulletin 326, 1972.

Petermann, W. G., Leaverton, E. D., and Ryker, N. L., "Alum Treatment of the Slopes of a 500-Foot Segment of Squaw Creek near Lawton, Oklahoma". Unpublished Report, Soil Conservation Service, USDA, Stillwater, Oklahoma, November 1975.

Ryker, N. L., "Encountering Dispersive Clays on Soil Conservation Service Projects in Oklahoma", American Society for Testing and Materials STP 623, 1976.

Sherard, J. L., Dunnigan, L. P., Decker, R. S., "Identification and Nature of Dispersive Soil", Journal of the Geotechnical Engineering Division, American Society of Civil Engineers, April 1975.

Sherard, J. L., Dunnigan, L. P., Decker, R. S. and Steele, E. F., "Pinhole Test for Identifying Dispersive Soils,", Journal of the Geotechnical Engineering Division, American Society of Civil Engineers, January 1976.

Dispersive Clay Erosion at Grenada Dam Mississippi

Edward B. Perry*
Member, ASCE

Abstract

One of the largest known areas of dispersive clays in the United States is in north-central Mississippi. Grenada Dam, constructed by the U.S. Army Engineer District, Vicksburg, lies in this area. Rainfall erosion tunnels developed on the downstream slope of the dam (protected by topsoil and Bermuda grass) soon after completion of the main embankment in 1949 and continued to develop at a decreasing rate to the present. Based upon laboratory tests conducted on undisturbed soil samples and reservoir water samples obtained in 1973 and 1976, the embankment soil is nondispersive at the surface and dispersive below a depth of about 6 ft. Limited data obtained indicate the foundation soil is dispersive at the surface and nondispersive to dispersive with depth. Based upon the history of occurrence of rainfall erosion tunnels on the downstream slope of the dam, it is apparent that the embankment soil at the time of construction was dispersive from the surface (below the topsoil) throughout its depth. The downstream slope of Grenada Dam has low susceptibility to rainfall erosion at present due to an amelioration process that has taken place with time and changed the upper portion of the embankment from a dispersive soil to a nondispersive soil. The Australian method of analysis using the exchangeable sodium percentage of the soil, total ionic concentration of the reservoir (eroding) water, and predominate clay mineral of the soil, indicate both the embankment and foundation soils would be potentially susceptible to dispersive clay piping if Grenada Dam contained cracks or sandy lenses traversing the width of the dam where the reservoir water would have a path of rapid access across the dam. Since piping failure through the embankment or foundation has not occurred, either the dam is free from cracks or sandy lenses traversing the width of the dam, or if cracks are present, the soil is able to swell and seal the flow channels.

Introduction

Dispersive clays are a particular type of soil in which the clay fraction erodes in the presence of water by a process of deflocculation. This occurs when the interparticle forces of repulsion exceed those of attraction so that the clay particles go into suspension and, if the water is flowing such as in a crack in an earth embankment, the detached particles are carried away and piping occurs.

*Research Civil Engineer, Soil Mechanics Division, Geotechnical Laboratory, U.S. Army Engineer Waterways Experiment Station, Vicksburg, MS 39180.

In addition to the possibility of piping failure, the slopes of earth embankments constructed with dispersive clays are susceptible to rainfall erosion. In natural soil deposits of dispersive clay, erosion due to wave action may occur along shorelines of reservoirs, and erosion due to current action may occur along channels constructed in dispersive clay (Sherard, Dunningan, and Decker 1977).

One of the largest known areas of dispersive clays in the United States is north-central Mississippi (Sherard 1972). Grenada Dam, which is located in this area, has had numerous subsurface erosion channels and subsequent cave-ins develop at relatively shallow depth on the downstream slope of the dam since construction of the main portion of the embankment in 1949 (Perry 1979).

The objective of this study was to determine the dispersion characteristics of embankment and foundation soils at Grenada Dam and susceptibility of these soils to rainfall erosion and dispersive clay piping, respectively.

Background Information on Dam

The Grenada Lake project is located on the Yalobusha River approximately 3 miles northeast of Grenada, Mississippi, and is part of a comprehensive flood control plan for the Yazoo River. The rolled earthfill dam is 13,900 ft in length with a crown width of 40 ft, and the maximum height above the valley floor is 80 ft. Figure 1 shows the typical cross sections of the dam. The upstream slope of the embankment is protected by a 2.5 - ft thickness of dumped riprap overlying a 1- ft - thick layer of gravel bedding. The downstream slope was covered with a 12-in. layer of topsoil and spot-sodded with Bermudagrass. Grenada Dam and appurtenances were constructed under two major contracts. The first contract covered construction of the main portion of the embankment, spillway, and most of the relief wells. Work on this contract was initiated in February 1947 and completed in September 1949. The second contract, delayed due to lack of funds, was awarded in April 1951 and provided for construction of the outlet structures, river closure section, and closure section for the highway, railroad, and power line. All work was completed and reservoir impounding began in January 1954.

Foundation and Borrow Materials

The foundation soils, with the exception of the Recent alluvium in the Yalobusha River floodplain, are of Tertiary origin, dating back to the Claiborne (middle Eocene) age. The Claiborne sediments consist of Basic City deposits of gray silts and clays with layers of shale and older underlying Meridian gray sand with occasional and discontinuous layers of silty sand and silty clay as shown in Figure 2. In May 1960, about 11 years after the completion of the main portion of the embankment, erosion patterns were documented in the excavated slope of borrow area DS as shown in Figure 3 (Perry 1979). This erosion pattern has been reported subsequently in excavated slopes of dispersive clay in Mississippi (Sherard, Decker, Ryker 1972).

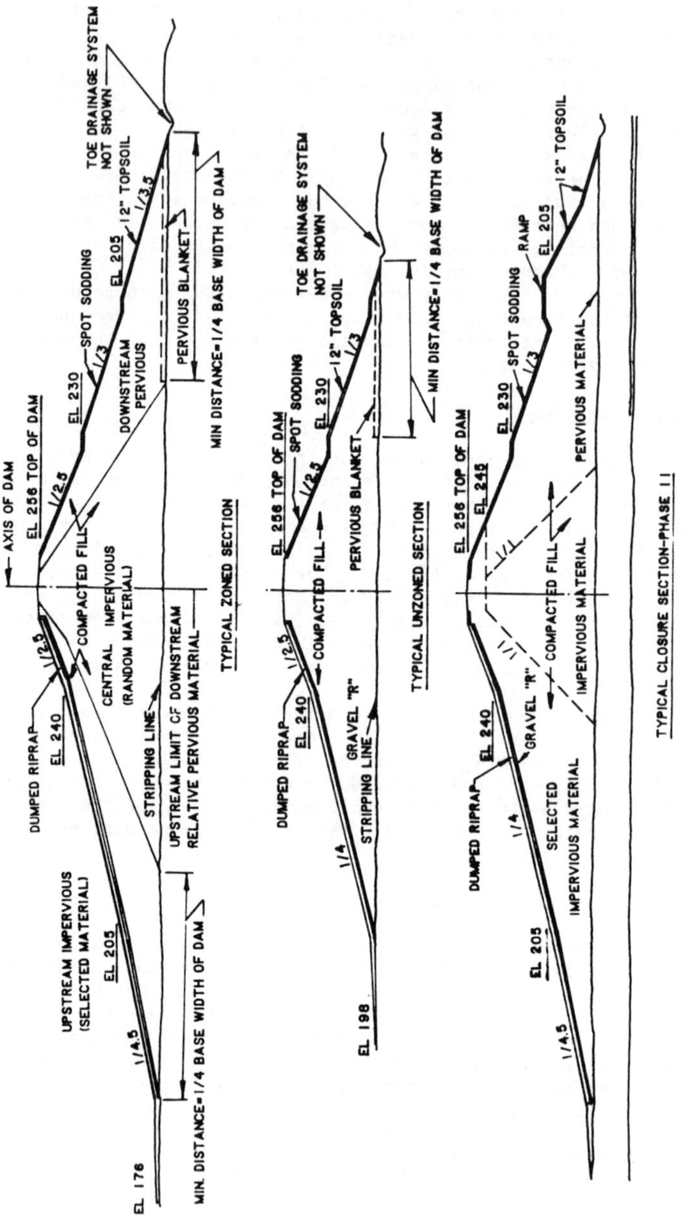

Figure 1. Typical cross sections of dam

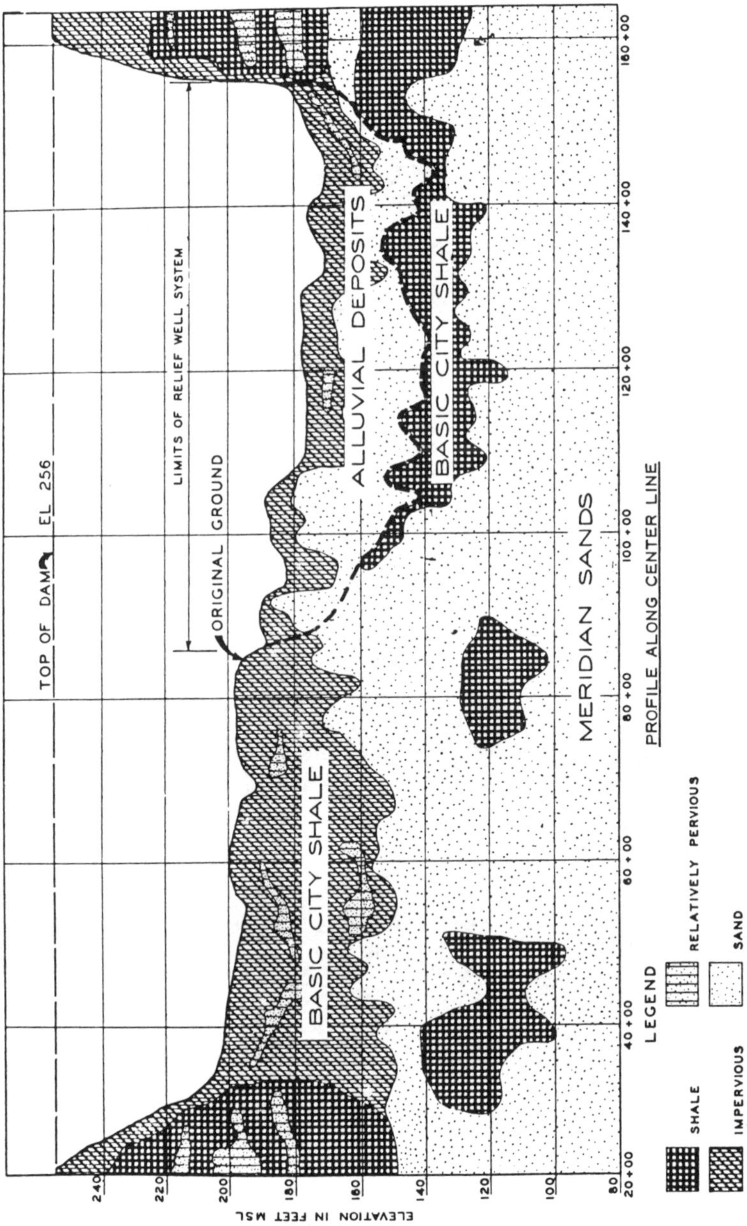

Figure 2. Generalized geological profile along center line of dam

a. General view of slope showing rills

b. Close-up view of slope showing vertical erosion tunnels and surface cracking resembling a karst topography

Figure 3. Erosion pattern of excavated slope of borrow area DS, 24 May 1960

Rainfall Erosion Tunnels on Downstream Slope

Rainfall erosion tunnels developed in the downstream slope of the dam soon after completion of the main embankment in 1949. The silty clay material in which the rainfall erosion tunnels developed was obtained from both the Tertiary terrace deposits and the Recent alluvium Yalobusha River floodplain deposits. Figure 4 shows schematic details of rainfall erosion tunnels. The terms "cave-in" and "jug" have been used to describe the vertical portion of the rainfall erosion tunnel. Although Figure 4 shows only one horizontal erosion outlet tunnel, there were often several horizontal erosion outlet tunnels for each cave-in.

The rainfall erosion tunnels are initiated when shrinkage cracks, as much as 1 in. wide at the ground surface and tapering to zero at a maximum depth of 7 ft, are formed during periods of low rainfall and high evaporation. At Grenada, Mississippi, from May through October the mean precipitation is about 19 in. while the mean pan evaporation is about 39 in. The type and amount of vegetation influence the amount of water lost through evapotranspiration, size and number of shrinkage cracks, and rainfall infiltration. Surface water from rainfall enters the cracks and erodes the soil (carries away the dispersed fines) before the cracks can swell and close. The cracks generally run together on the surface of the slope in a dendritic pattern, like the branches of a tree, to form progressively larger cracks as the surface water moves down the slope. Eventually the vertical portion of the rainfall erosion tunnel becomes so large that the overlying soil collapses into the cavity. Although the development of the rainfall erosion tunnels with time was not investigated, it has been suggested that the cave-ins develop over a period of several years. Although the dam was not damaged structurally by the rainfall erosion tunnels, which were limited to a depth of about 7 ft, the cave-ins were unsightly, some manholes along the collector pipe for the toe drainage system (prior to its replacement with an open paved ditch in 1961) were undermined, and the stability of some concrete drop inlets located on downstream berms was threatened. Repair measures to the downstream slope consisted of excavating the rainfall erosion tunnels, recompacting the same material in the channels with power tampers, and sodding.

During the winter months from October 1955 to January 1956, the rainfall was 9.8 in. below normal as shown in Figure 5. Prior to 3 February 1956, there were no cave-ins noticed during an inspection of the downstream slope. From 1-6 February 1956, a total of 5.6 in. of rainfall occurred as shown in Figure 6. An inspection of the downstream slope on 7 February revealed 55 cave-ins and numerous horizontal outlet tunnels. During the period 16-20 February 1956, a total of 2.28 in. of rainfall occurred, after which it was found that the number of cave-ins had increased to 114. On 14 March 1956, there were 133 cave-ins and 459 horizontal outlet tunnels found on the downstream slope. Figure 7 shows the number of cave-ins versus time beginning with the year 1956 when records became available. Over 100 cave-ins per year were recorded for the period 1956 to 1960. Gradually with time the number of cave-ins diminished to about 5 cave-ins per year for the past 10 years.

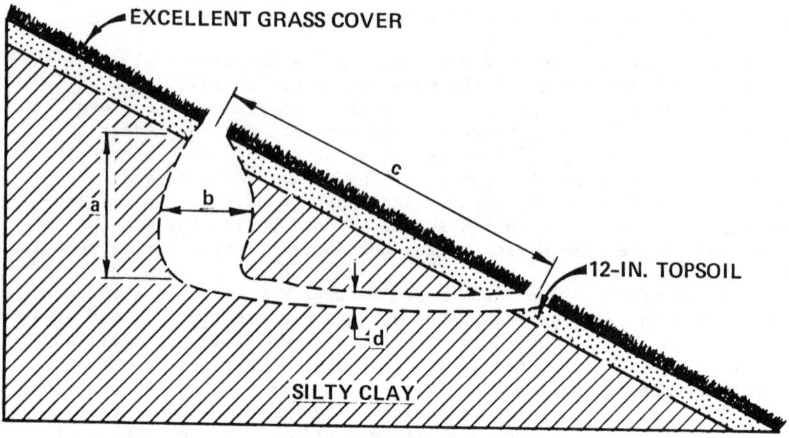

	DIMENSIONS, FT		
	MINIMUM	AVERAGE	MAXIMUM
a.	1	4	7
b.	1	3	5
c.	2	6	25
d.	0.5	1	3

Figure 4. Schematic details of rainfall erosion tunnel in downstream slope

Laboratory Soils Testing

Undisturbed soil samples taken by the Vicksburg District were tested at the WES in 1976 to determine the dispersive characteristics of embankment and foundation soils. Test conducted included classification and index tests, soil (and reservoir water) chemistry tests, and dispersion tests [Crumb, Soil Conservation Service (SCS) dispersion, soil pore water chemistry correlation, and pinhole erosion tests]. Most of the embankment soil was classified as lean or sandy clays (CL) with a few soil samples classified as plastic or sandy clay (CH) and clayey sand (SC). The foundation soil was classified as lean, sandy, or silty clay (CL) and clayey sand (SC). Using the consensus of the SCS dispersion test results, soil pore water chemistry correlation, and pinhole erosion test results, a summary of dispersion versus depth is given in Figures 8-10. The horizontal scale in Figures 8-10 represents a consensus of test results from the SCS dispersion test (0-35% nondispersive, 35-50% intermediate, and 50-100%

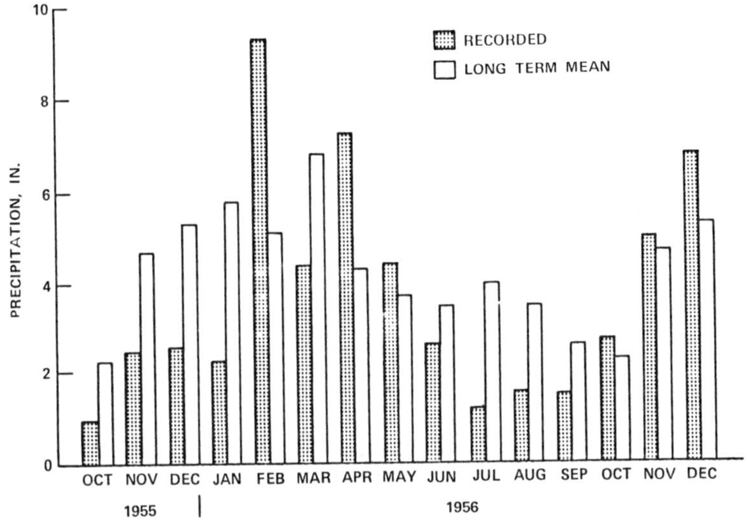

Figure 5. Precipitation from October 1955 to December 1956 for Grenada, Mississippi

dispersive); soil pore water chemistry correlation, (nondispersive, intermediate, and dispersive), and pinhole erosion test (ND_1 - ND_2 nondispersive, ND_3 - ND_4 intermediate, and D_1 - D_2 dispersive). The embankment soil is nondispersive at the surface and dispersive below a depth of about 6 ft (Perry 1979).

Susceptibility of Embankment Soil to Rainfall Erosion

Based upon the history of occurrence of rainfall erosion tunnels on the downstream slope of the dam, given in Figure 7, apparently the embankment soil at the time of construction was dispersive from the surface (below the 12-in. layer of topsoil) throughout its depth. Based upon the results of this study, the downstream slope of Grenada Dam has low susceptibility to rainfall erosion at present due to an amelioration process that has taken place with time that has changed the upper portion of the embankment from a dispersive soil to a

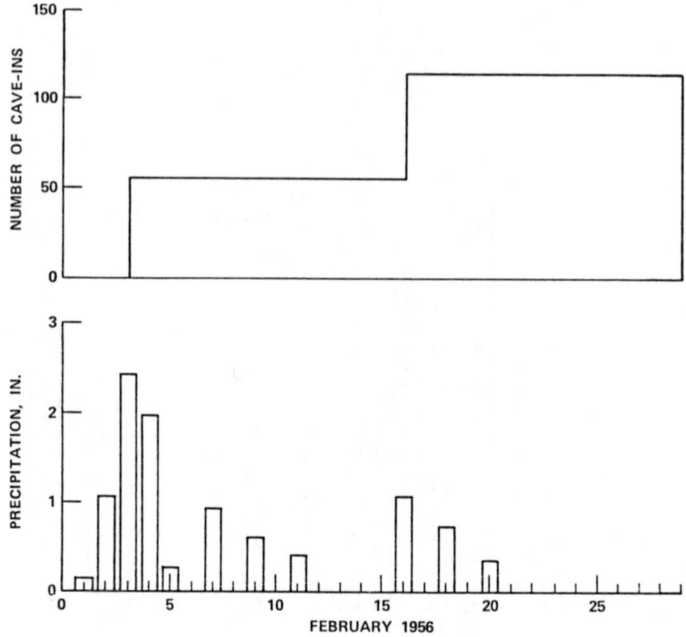

Figure 6. Precipitation and cave-ins during February 1956

nondispersive soil. Changes in the dispersion characteristics of the upper portion of the embankment may result from seasonal moisture and temperature variations, biological activity, and root growth.

Susceptibility of Embankment and Foundation Soils to Piping

Australian workers have developed a method of analysis using the exchangeable sodium percentage (ESP)* of the soil, total ionic concentration of the eroding water, and predominate clay minerals of the soil to assess the susceptibility of a homogeneous earth dam to dispersive clay piping (Ingles and Metcalf 1973). If the dam contains cracks or sandy lenses traversing the width of the dam, the reservoir

---

$$*ESP = \frac{\overline{Na}}{\overline{CEC}} (100)$$

where $\overline{NA}$ = exchangeable sodium

$\overline{CEC}$ = cation exchange capacity

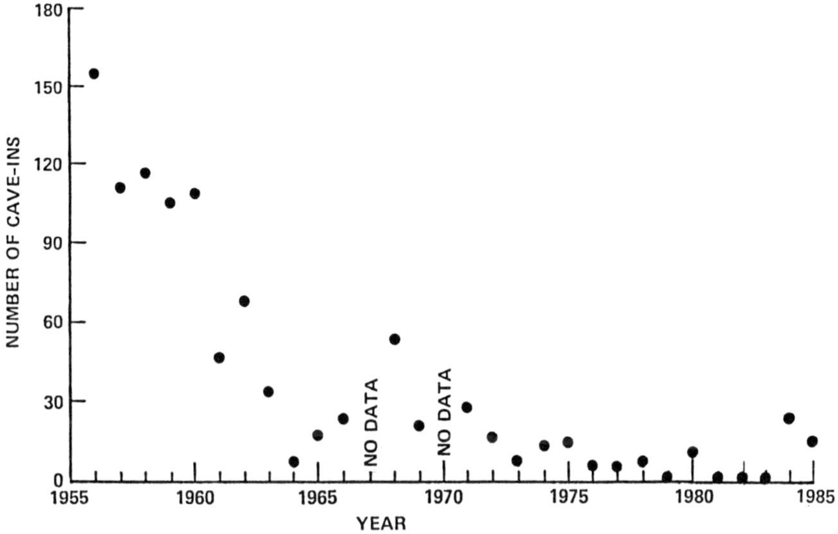

Figure 7. Number of cave-ins, 1956-1985

(eroding) water will have a path of rapid access across the dam, and the dam may fail by dispersive clay piping upon first filling of the reservoir. The only deterrence to failure by dispersive clay piping for a homogeneous dam would be the ability of the embankment material to swell and seal the flow channels. Using the ESP of the soil, total ionic concentration of the reservoir (eroding) water, and predominate clay mineral of the soil summarized in Table 1, the susceptibility of Grenada Dam to dispersive clay piping, assuming the reservoir water has a path of rapid access across the dam, is shown in Figure 11. Since no appreciable changes in the ESP of the soil, total concentration of the reservoir water, or predominate clay mineral of the soil are believed to have occurred from impoundment of the reservoir in 1954 to present, Figure 11 would be applicable for any point in time. With the Australian method of analysis, the plot of both the embankment and foundation soils in the zone of potential deflocculation (see Figure 11) indicates that if Grenada Dam contained cracks or sandy lenses traversing the width of the dam where the reservoir (eroding) water would have a path of rapid access across the dam, there would be a potential for dispersive clay piping. Since piping failure through the embankment or foundation has not occurred at Grenada Dam, either the dam is free from cracks or sandy lenses traversing the width of

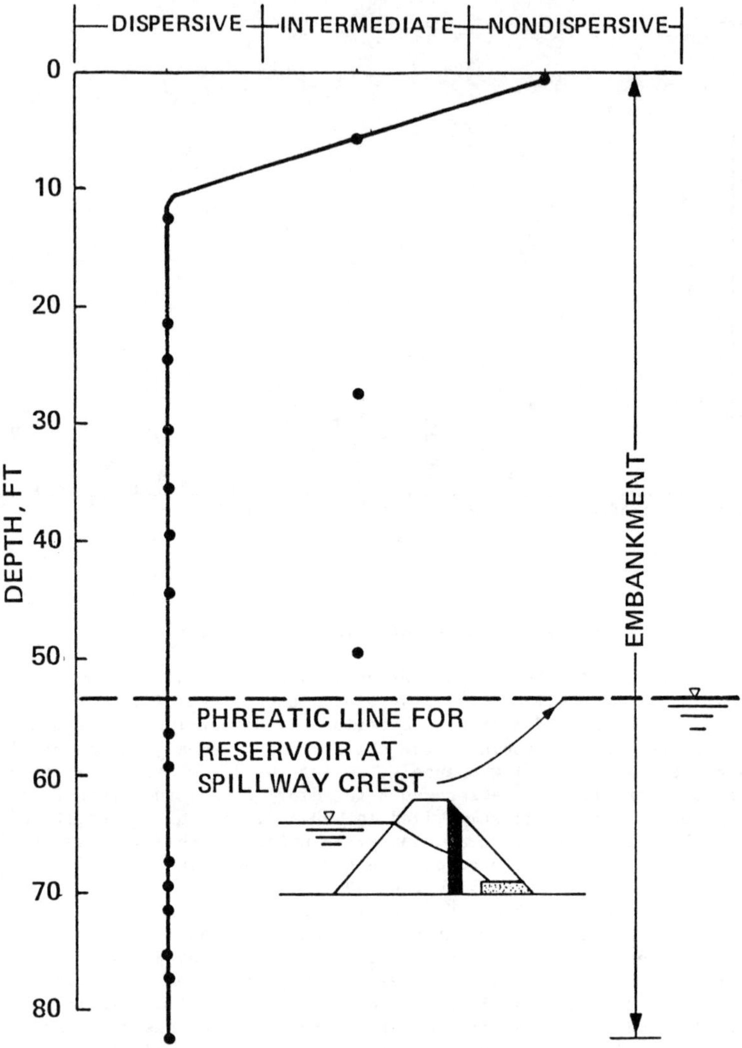

Figure 8. Summary of dispersion versus depth for boring 1

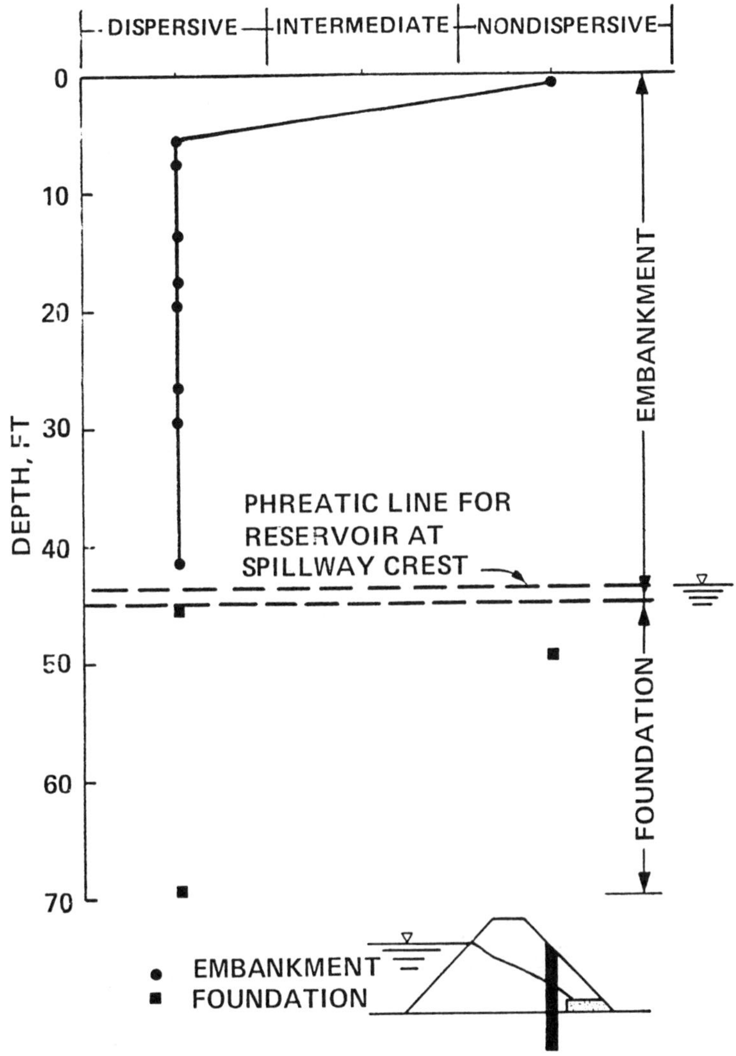

Figure 9. Summary of dispersion versus depth for boring 2

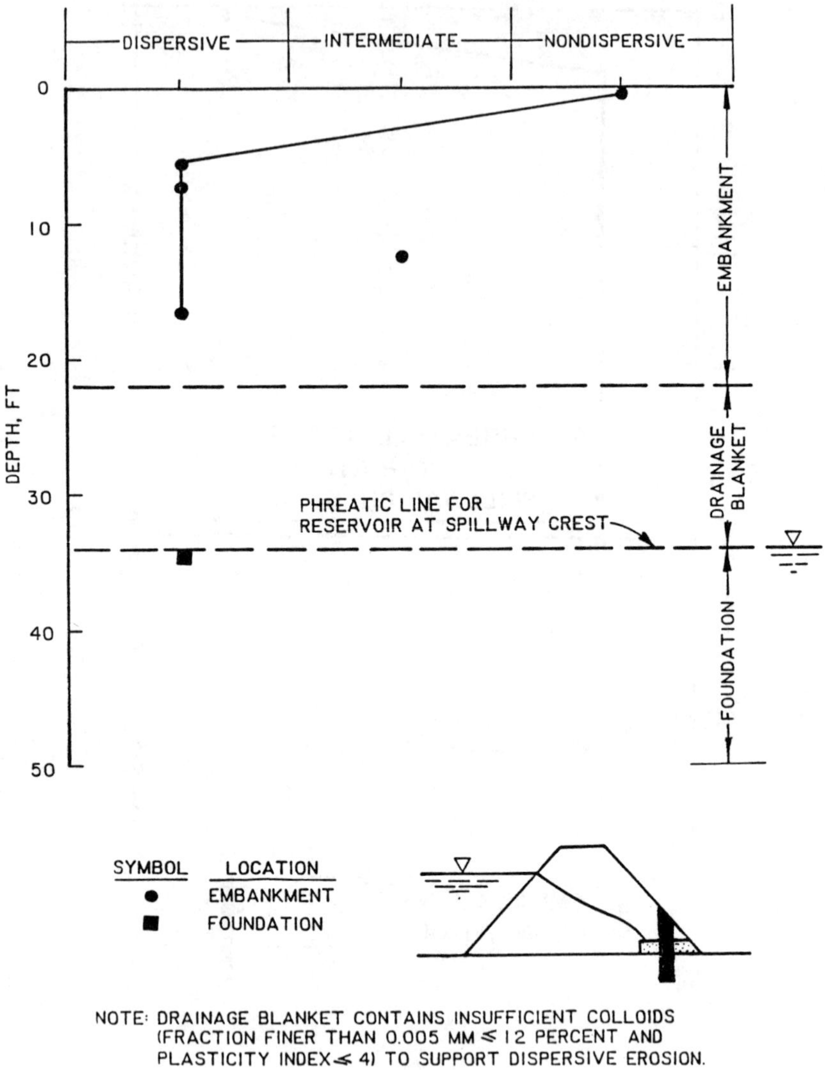

Figure 10. Summary of dispersion versus depth for boring 3

the dam, or if cracks are present, the soil is able to swell and seal the flow channels.

TABLE 1.--Summary of Parameters for Prediction of Susceptibility of Embankment and Foundation Soils to Piping Failure

Boring No.	Location	Soil: ESP*	Soil: Predominate Clay Mineral	Reservoir (Eroding) Water Total Ionic Concentration meq/l
1	Embankment	12	Montmorillonite	0.6
2	Embankment	17	Montmorillonite	0.6
3	Embankment	16	Montmorillonite	0.6
2	Foundation	6	Montmorillonite	0.6
3	Foundation	16	Montmorillonite	0.6

$*ESP = \frac{\overline{Na}}{CEC}(100)$

Conclusions

Grenada Dam was constructed from 1947 to 1954 in north-central Mississippi in what is now known to be one of the largest areas of dispersive clay in the United States. Numerous rainfall erosion tunnels developed on the downstream slope of the dam following completion of the main embankment and continued to develop at a decreasing rate to the present. The downstream slope of Grenada Dam has low susceptibility to rainfall erosion at present due to an amelioration process that has taken place with time and changed the upper portion of the embankment from a dispersive soil to a nondispersive soil. The Australian method of analysis using the exchangeable sodium percentage of the soil, total ionic concentration of the reservoir (eroding) water, and predominate clay mineral of the soil, indicate both the embankment and foundation soils would be potentially susceptible to dispersive clay piping if Grenada Dam contained cracks or sandy lenses traversing the width of the dam where the reservoir water would have a path of rapid access across the dam. Since piping through the embankment or foundation has not occurred, either the dam is free from cracks or sandy lenses traversing the width of the dam, or if cracks are present, the soil is able to swell and seal the flow channels.

Acknowledgements

This study was conducted by the U.S. Army Engineer Waterways Experiment Station for the U.S. Army Engineer Lower Mississippi Valley Division. This study was planned by Dr. E. B. Perry under the general supervision of C. L. McAnear, Chief, Soil Mechanics Division, and Dr. W. F. Marcuson, III, Chief, Geotechnical Laboratory (GL). Undisturbed soil samples from the embankment and foundation of Grenada Dam were furnished by the Vicksburg District. Reservoir water samples from Grenada Lake were obtained through the cooperation of

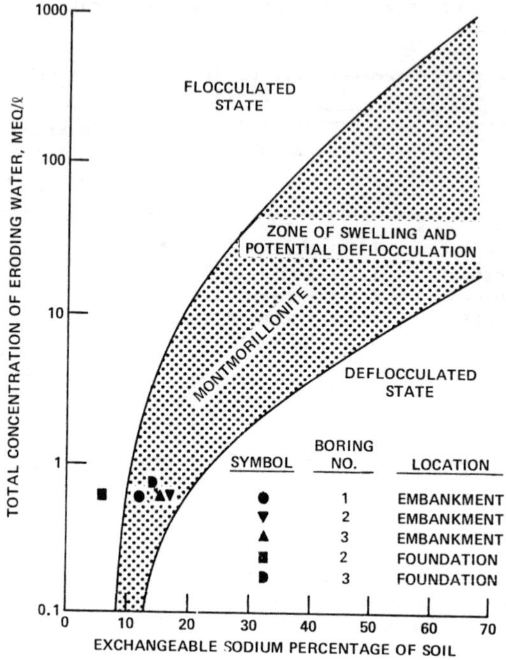

Figure 11. Prediction of piping failure using the total ionic concentration of the eroding water and the ESP of the soil assuming the reservoir water has a path of rapid access across the dam (adapted from Ingles and Metcalf 1973)

F. Methvin, Jr., Assistant Resource Manager, Grenada Lake Field Office. Pinhole erosion tests were conducted by L. R. Coffin, Jr., under the direct supervision of G. P. Hale, Chief Soils Research Center, GL. Soil and reservoir water chemistry tests were conducted by D. L. Bean under the general supervision of T. B. Husbands, Chief, Chemistry Unit, Structures Laboratory. Permission was granted by the Chief of Engineers to publish this information.

Appendix.- References

Ingles, O. G., and Metcalf, J. B., Soil Stabilization, John Wiley and Sons, Inc., New York, N.Y., 1973.

Perry, E. B., "Susceptibility of Dispersive Clay at Grenada Dam, Mississippi, to Piping and Rainfall Erosion", Technical Report GL-79-14, U.S. Army Engineer Waterways Experiment Station, Vicksburg, Miss., Sept. 1979.

Sherard, J. L., "Study of Piping Failures and Eroding Damage from Rain in Clay Dams in Oklahoma and Mississippi", U.S. Department of Agriculture Soil Conservation Service, Washington, D. C., Mar. 1972.

Sherard, J. L., Decker, R. S., and Ryker, N. L., "Piping in Earth Dams of Dispersive Clay", Proceedings, Speciality Conference on Performance of Earth and Earth-Supported Structures, ASCE, Vol 1, Part 1, 1972, pp. 589-626.

Sherard, J. L., Dunnigan, L. P., and Decker, R. S., "Some Engineering Problems with Dispersive Clays", Dispersive Clays, Related Piping, and Erosion in Geotechnical Projects, ASTM, Special Technical Publication No. 623, May 1977, pp. 3-12.

Piping Susceptibility and Filter Criteria for Sands

F. C. Townsend,[1] M. ASCE, Jih-Min Shiau[2] and T. J. Pietrus[2]

ABSTRACT

A laboratory flume with a transparent plastic top and filled with sand was used to investigate piping and filter criteria by subjecting the saturated sand specimen to variable pressure heads. With small initial pipes formed in the sand, the critical heads (gradients) required to initiate piping in six different sand gradations were determined. It was determined that piping requires a discontinuity as an initiator to concentrate vertical gradients, which, coupled with horizontal gradients, create erosion. However, meandering of the pipe could lead to decreased velocities, resulting in downstream deposition and self-healing tendencies. Gradation effect studies showed uniform fine grained sands are more susceptible to piping than coarse grained well-graded sands.

Current COE and SCS filter criteria, $D_{15}/d_{85} \leq 5$ or ≤ 4 have a safety factor of about two. Also the NHSRI criteria for internal stability can be used to assess the piping susceptibility of well- and gap- graded sands.

INTRODUCTION

A laboratory testing program was undertaken in order to understand better the phenomena of hydraulic piping in sands and to evaluate current filter design criteria. This research was initiated following the failure of a 35 ft. (11.5 m) high homogeneous sand dike under a low head used to retain cooling water for a power plant in Florida. Accordingly, our objectives were to: (a) reproduce piping of sands in the laboratory (b) examine hydaulic and sand characteristics most conducive to piping, and (c) evaluate filter criteria for mitigating piping.

To accomplish these objectives, a 1 ft. x 1 ft. x 7.5 ft. (30 x 30 x 229 cm) laboratory flume with a transparent plastic top was constructed to provide visual observation of piping. Six sand gradations: three uniform gradations, one well graded gradation, and two gap graded gradations, were tested for piping susceptibility. Using the most piping susceptible sand, three filter design gradations and filter thicknesses subsequently were evaluated.

[1] Prof., Dept. of Civil Engr., Univ. of FL., Gainesville, FL 32611
[2] Grad. Res. Asst., Dept. of Civil Engr., Univ. of Fl., Gainesville, FL 32611

BACKGROUND

Piping - "Piping" is the phenomenon of internal soil erosion within an embankment dam or foundation and is a major cause of unfavorable dam incidents, i.e., Teton Dam (Seed and Duncan 1981), Fontinelle Dam (Bellport, 1967). The pipe appears to originate with cracks or joints which serve to concentrate flow at the pipe tip. If the hydraulic gradients become sufficiently large at the tip, internal piping erosion commences and will progress upstream. Usually, the pipe is assumed to continue to migrate upstream at an accelerating rate, due to the increased hydraulic gradient, provided there is sufficient flow to transport the eroded material out through the pipe.

The seepage velocities required for erosion of various particle diameters can be estimated from Shield's equation for horizontal stream beds in the absence of seepage forces. By equating the average wall shear stress, τ_o, to Shield's equation for critical shear stress, τ_{cr} (Graf, 1971) one obtains:

$$\tau_{cr} = 0.06\ (\gamma_s - \gamma_w)\ de \text{ and } \tau_o = (\gamma_w\ RV_m^2)/[M^2\ (R)^{4/3} \approx, \text{ gm/cm}^2$$

where γ_s, γ_w = unit weight of solids or water, respectively

and $(\gamma_s - \gamma_w) = 1.7\gamma_w$, gm/cm^3

de = effective grain size, d_{35}, cm

R = hydraulic radius (R = d/4, circular pipe)

V_m = Seepage velocity cm/sec = discharge velocity ÷ porosity

M = $8.25\ \sqrt{g}/(de)^{1/6}$, g = gravity

d = dia of pipe, cm

the result being:

$$V_m = 65.47\ de^{1/3}\ d^{1/6}, \text{ cm/sec.}$$

If one considers a fine sand particle (de = 0.2 mm) and a very small pipe (d = 1.0 cm), the velocity required for erosion is approximately 18 cm/sec. Obviously, this velocity far exceeds those occurring through soils and some other factor is required for initiating erosion in a pipe.

Richart and Schmertmann (1958) theorized that a significant vertical gradient into a horizontal cohesionless soil surface can greatly reduce the hoizontal velocity required for scour. Because the horizontal velocities at the leading edge of a pipe are quite small, it was hypothesized that the vertical gradient required to cause erosion must approach the critical gradient for quick conditions. Thus, a discontinuity (crack, joint, etc.) must exist to concentrate sufficient flows to initiation erosion.

FILTER DESIGN

Traditional filter design has relied upon the U.S. Army Corps of Engineers (COE) criteria as presented in Table 1. More recently, the Soil Conservation Service (SCS) [Sherard, et. al., (1984), and Sherard and Dunnigan (1985)] has offered filter criteria for sands. The SCS criteria, for which Group 3 soils are germaine to this discussion, are presented in Table 2 and are based upon D_{15} particle

TABLE 1

FILTER DESIGN CRITERIA OF THE U.S. ARMY CORPS ENGINEERS

Criteria (1)	Designation (2)
$D_{15f}/d_{85b} < 5$	piping ratio
$5 < D_{15f}/d_{15b} < 20$	permeability ratio
$D_{50f}/d_{50b} < 25$	parallel gradation

D = Diameter of percent finer for filter material

d = Diameter of percent finer for base material

TABLE 2
FILTER DESIGN CRITERIA OF THE SOIL CONSERVATION SERVICE[a]

Soil group (1)	% Fines (<No. 200 sieve) (2)	Recommended design criteria (already includes adequate safety factor) (3)
1	85 to 100%	$D_{15}/d_{85} \leqslant 9$
2	40 to 85%	$D_{15} \leqslant 0.7$ mm
3	0 to 15%	$D_{15}/d_{85} \leqslant 4$
4	15 to 40%	Intermediate between group 2 and 3

a (From Sherard & Dunnigan, 1985)

size. Sherard, et. al, (1984) pointed out that the internal pore channels are governed by the finer D_{15} sizes and that since the rationale for filter criteria employing D_{50}, d_{15} and d_{50} sizes is not satisfactory, only D_{15} criteria are used. For Group 3 soils, the criterion $D_{15} \leqslant 4\ d_{85}$ is comparable to the COE $D_{15} \leqslant 5d_{85}$ criterion. Sherard and Dunnigan (1985) suggest that laboratory tests exhibit failure at about $D_{15}/d_{85} = 7.5$ to 9, hence the $D_{15} \leqslant 4\ d_{85}$ criterion has a safety factor of about two.

The Northwest Hydrotechnical Science Research Institute (NHSRI), China, also proposed filter design criteria for cohesionless soils. To use these filter criteria, the internal stability of the soil to resist particle migration is considered as presented in Table 3. Based upon gradation, uniformity coefficient (C_u), the portion of fine material in gap-graded soils, and the average pore diameter (D_o), soils are classified as stable, unstable, or transitive. Uniform soils ($C_u \leqslant 5$) are considered as being internally stable. For non-uniform soils ($C_u > 5$), a distinction is made between gap- and continuous-gradations. Stable gap-graded soils are those containing more than 35% (>35%) as the finer portion of the gap grading. Stable well-graded soils are those whose average pore diameter, D_o, is less than D_3 ($D_o < D_3$); where D_o can be approximated as $D_o \approx 0.63\ nD_{20}$ (n = porosity, D_{20} = 20 percent finer grain diameter), and D_3 = 3% finer grain diameter.

The NHSRI filter criteria presented in Table 4 are based upon the uniformity of the base soil, the internal stability of the base soil, and the ratio D_{20}/d_k, where d_k is the control grain diameter of the base soil as presented in Table 5 and Figure 1.

LABORATORY EQUIPMENT AND TESTING PROCEDURES

The laboratory tests were performed in a 7.5 ft. x 1 ft. x 1 ft. (229 x 30 x 30 cm) aluminum flume with a transparent plastic top covering the sand as presented in Figure 2. An internal weir, used for forming a downstream submerged slope, provided a test length of 5 ft. (152 cm). A 1 in. (2.54 cm) thick plexiglass top sealed with an "O" ring acts as an impermeable roof through which piping can be observed. A rubber bladder at the bottom of the flume provides an upward pressure when inflated that maintains the sand in contact with the plastic top. Manometers placed at quarter points and the ends provide measurement of total heads acting along the sand. On the downstream end of the flume, an access well was constructed through which the specimen can be reached.

The testing procedure consisted of the following steps; (a) specimen preparation, (b) creation of initial pipes, (c) filter construction, (d) vacuum and saturation and (e) testing.

TABLE 3
NHSRI CLASSIFICATION OF COHESIONLESS SOILS FOR INTERNAL STABILITY

Soil	Uniformity	Gradation	Criterion	Classification
Cohesionless soil	Uniform Soil ($C_u<5$)			Stable soil
	Non-Uniform ($C_u>5$)	Gap gradation	$P > 35\%$	Stable soil
			$P < 25\%$	Unstable soil
			$P = 25\text{–}35\%$	Transitive soil
		Continuous gradation	$D_o < D_3$	Stable Soil
			$D_o > D_5$	Unstable soil
			$D_o = D_3\text{-}D_5$	Transitive soil

P = THE PORTION OF FINER MATERIAL IN GAP GRADATION

D_o = THE AVERAGE DIAMETER OF PORES, $D_o = 0.63nD_{20}$ (N = POROSITY)

(From Liu, 1984)

TABLE 4
FILTER DESIGN CRITERIA
FOR COHESIONLESS SOILS PROPOSED BY NHSRI

		Design Criteria
Cu ≤ 5		$D_{20}/d_{70} \leq 7$
Cu > 5	stable soil[a]	$D_{20}/d_k \leq 7$
	unstable [a]	$D_{20}/d_{15} \leq 5$

d_k = the control grain diameter of protected soil (Table 5)

(a) Classification Table 3

TABLE 5
CONTROL GRAIN DIAMETER d_k

	Control grain diameter d_k		
	stable soil (a)		unstable soil (a)
	Cu ≤ 5	Cu > 5	
Continuous gradation	d_{70}	d_k(Figure 1)	d_{15}
Gap gradation		d_{70}xP	d_{15}

P = Percent of fine portion in gap gradation

(a) Classification Table 3

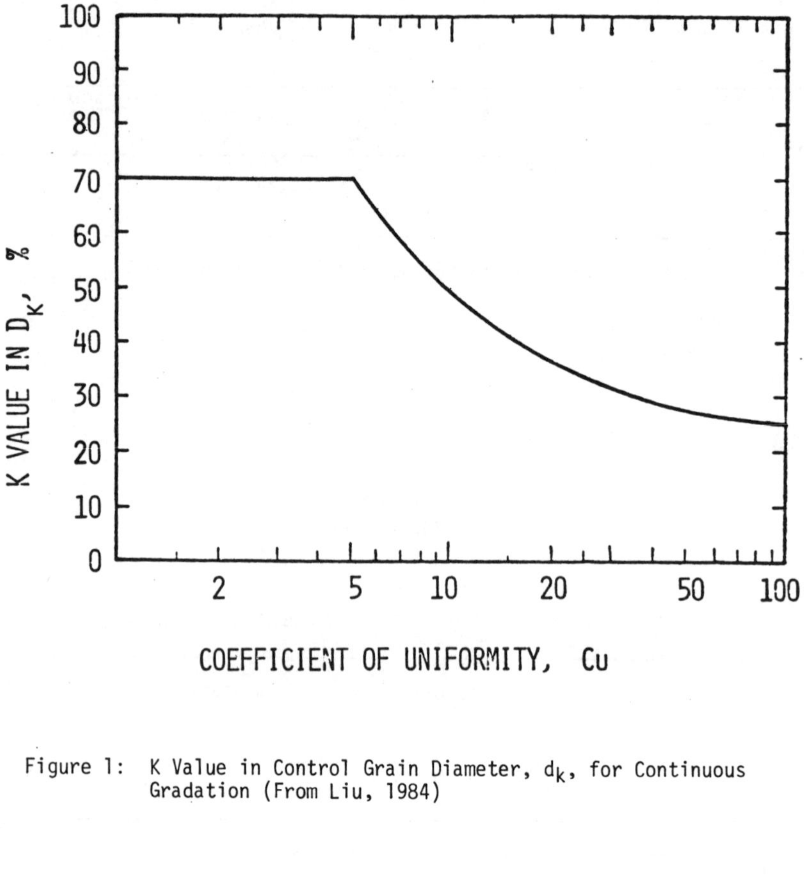

Figure 1: K Value in Control Grain Diameter, d_k, for Continuous Gradation (From Liu, 1984)

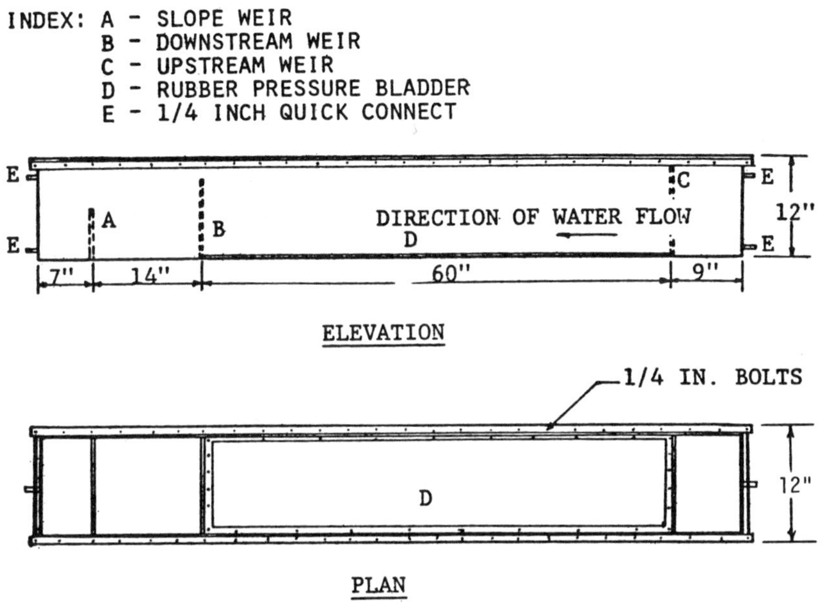

Figure 2: Hydraulic Flume Schematic
(1 in. = 2.54 cm)

Specimen preparation consisted of filling the flume between the upstream and downstream weir by raining the sand through 1/2 in. (1.3 cm) holes from a height of 3 to 2 ft. (91 to 61 cm) [the sand specimen is one ft (30 cm) thick] using a traversing hopper. Density determinations were periodically taken using buried tare cans of known volume. Lastly, the surface was screeded to obtain a smooth surface.

The initial pipes to concentrate flows were formed using semi-circular wooden dowels of varying diameters of 1/4, 1/2, or 1.25 in. (6, 13, or 32 mm) and penetration lengths of 10 to 50 percent of the 60 in. (152 cm) specimen length. In later tests, sugar which was dissolvable, was used to create the initial pipes

For tests involving filters, the filter materials were placed between two vertical cookie sheets supported by wooden dowels. The top surfaces were covered with plastic to prevent contamination during filling the flume with sand. After raining of the sand, the cookie sheets were pulled.

With the dowel or sugar in place, the plastic top was bolted to the flume to have a fully sealed condition. A bladder pressure of 5 psi (34 kPa) was applied to push the specimen into contact with the plastic lid. In early tests, a vacuum was applied to the specimen prior to seepage saturation, but later this was abandoned. After saturation, the pipe forming semi-circular dowel was gently pulled from the sand leaving an indentation for the initial pipe.

Testing consisted of incrementally increasing the constant head acting on the sand specimen either by raising a constant head reservoir, or in cases when higher flows were required, by adjusting a pressure regulator connected directly to city water. In both cases, the average head was known as the difference between the first upstream manometer and the downstream overflow weir. Visual observations easily noted when piping was initiated.

MATERIALS

Six sand gradations were used in this research to examine piping and filter design as shown in Figure 3. Three of the gradations were uniform: Reid Bedford, 20/30, and 8/30, but with different effective grain diameters, D_{10} = 0.14, 0.63, and 0.8 mm, respectively. One gradation was well graded, and the remaining two gradations gap-graded, but also with different D_{10} values, 0.16, and 0.28 mm. Table 6 summarizes the gradation of these sands.

Three filter gradations were used as presented in Figure 4 and as tabulated in Table 7. For all filter evaluation tests, the most piping susceptible sand, Reid Bedford, was the only base sand used.

PIPING PROCESS OBSERVATIONS

The importance of the laboratory model tests was the seldom possible opportunity to observe visually the formation and

Table 6
Grading Characteristics of Sands Tested

Sand (1)	D_{10} (mm) (2)	D_{50} (mm) (3)	C_u (4)	C_c (5)
Reid Bedford	0.14	0.20	1.5	1.23
20/30	0.63	0.93	1.6	0.97
8/30	0.8	1.6	2.1	1.48
Well graded	0.24	1.42	6.7	2.20
Gap I	0.16	0.50	5.6	0.31
Gap II	0.28	0.60	6.1	0.26

Table 7
Comparison Between Filter Materials and Design Criteria

Criteria [a]	Filter			Criteria	
	Filter I	Filter II	Filter III	COE	SCS [b]
(1)	(2)	(3)	(3)	(5)	(6)
D_{15}/d_{85}	2.8	8.0	12.0	< 5	≤ 4
D_{15}/d_{15}	5.2	14.7	22.1	$> 5, < 20$	-
D_{50}/d_{50}	7.6	16.8	21.2	< 25	-

(a) D = Filter mat'l, d = base mat'l

(b) Sherard & Dunnigan, 1985

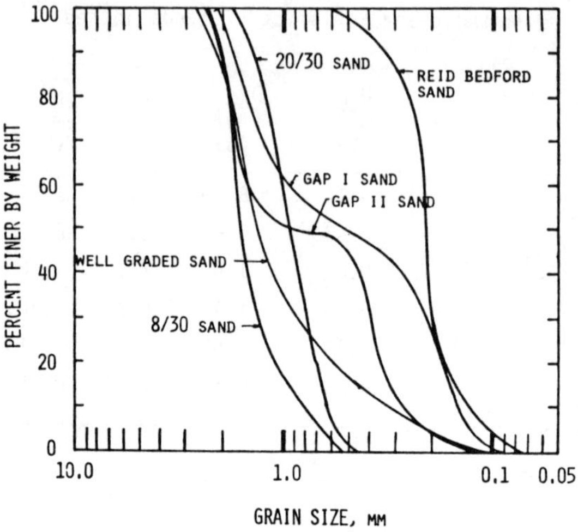

Figure 3: Grain Size Distribution of the Sands Tested

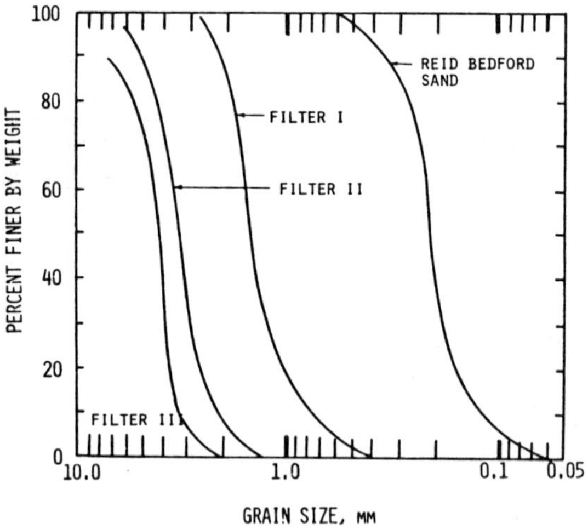

Figure 4: Grain Size Distribution of the Filter Materials

progression of a pipe. As the critical head (gradient) was achieved, the pipe began to form by the sloughing of several sand particles at the pipe tip into the void as illustrated in Figure 5. No "boiling" or such action occured, but this was expected as there is insufficient flow of water to sustain boiling, and because the vertical gradient need only approach the quick condition value so that the horizontal flow component causes erosion. Actually, what was observed was the pipe advancing via continuous miniature landslides progressing upstream with the displaced particles washed down the pipe. As the pipe developes, it follows the path of least resistance and meanders as it progresses upstream. Consequently, scouring occurs at some points while deposition occurs at others causing the pipe to vary from several sand grains wide to approximately 1/4-inch (6 mm) downstream. Figure 6 illustrates an example pipe.

In several cases, the increased width of the pipe caused by meandering reduced the flow sufficiently that transported particles deposited themselves in the pipe. This deposition eventually led to blockage causing cessation of the pipe and self-healing occurred. Thus, contrary to expectations that as the pipe traveled upstream pipe progression would accelerate due to increased gradients (Δl decreasing), piping retardation was observed. For these cases, the head was eventually increased until piping was re-initiated.

PIPING GRADIENTS

Once the critical heads necessary to initiate piping were obtained, the <u>average</u> horizontal gradient (i_{ave}) was easily calculated by <u>dividing</u> this head by the specimen length (60in., 152 cm). Table 8 presents the piping test results with critical heads (if any) obtained, the <u>average</u> gradients, and the final gradient required to maintain piping.

The effects of initial pipe penetration and diameter for the Reid Bedford sand tests are presented in Figure 7. These results show that as anticipated, smaller diameter pipes require lower gradients to initiate piping. This observation is consistent in that small diameter pipes concentrate flows more than large diameter pipes, thereby creating higher gradients and velocities to initiate erosion. The results also show that as the percent penetration increases, the average gradient, i_{ave}, to initiate piping decreases. This observation is consistent that for greater penetrations, the actual gradient is greater due to the reduced distance from the headwaters.

The effects of grain diameter on the piping susceptibility of uniform sands are presented in Figure 8. For uniform sands, the results show that, as anticipated, the smaller grain diameters are more susceptible to piping. Obviously, lower velocities are required to erode smaller particles.

The effects of gradation are summarized in Table 9, and show that the well-graded and gap-graded soils did not sustain piping,

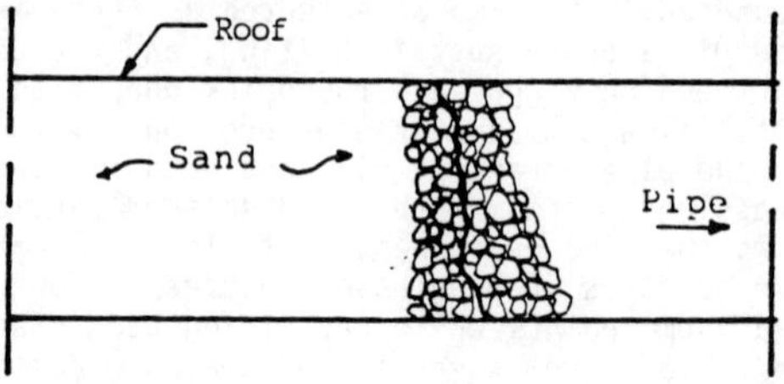

Section of material as pipe developes.

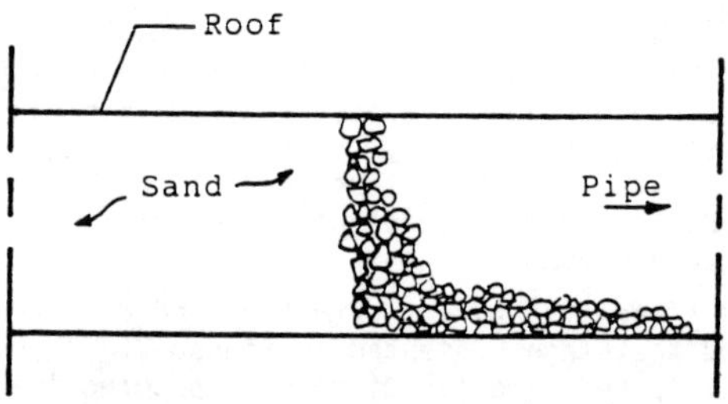

Note material after sliding into the pipe and carried downstream.

Figure 5: Development of Piping Action

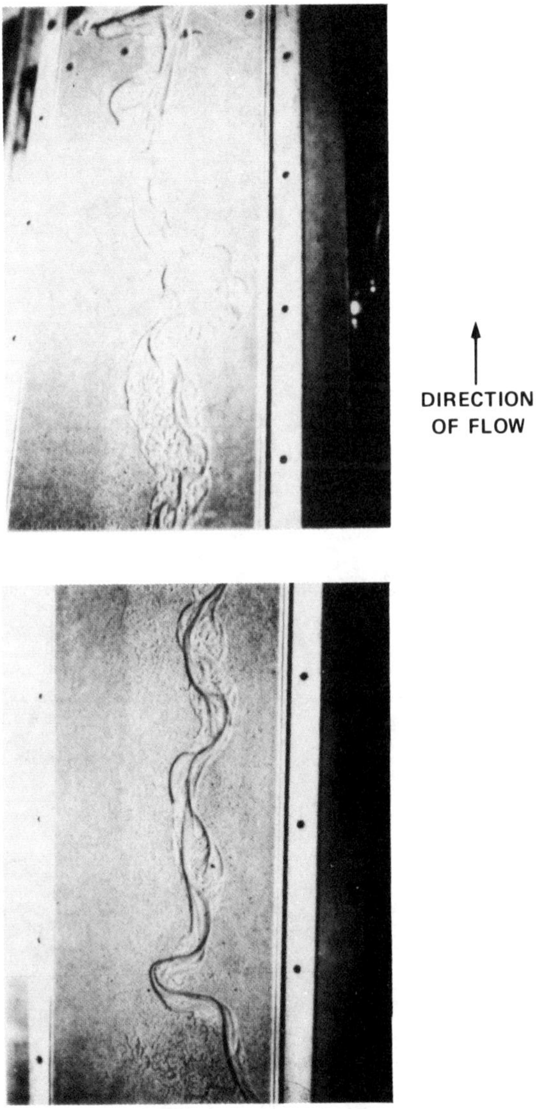

Figure 6: Close Up View of Pipe at Downstream (Top) And Upstream (Bottom) Ends

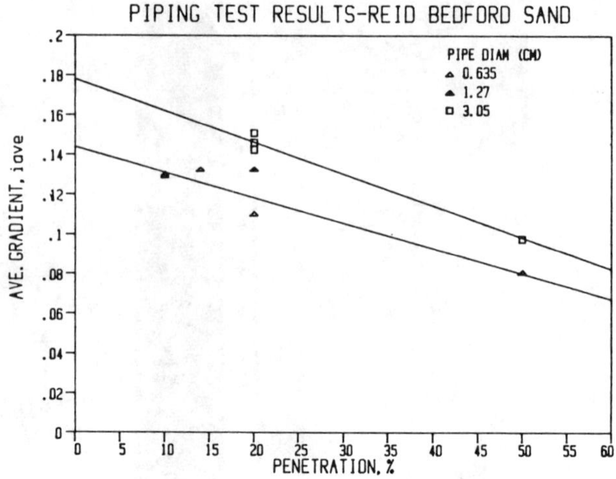

Figure 7: Effects of Pipe Penetration and Diameter

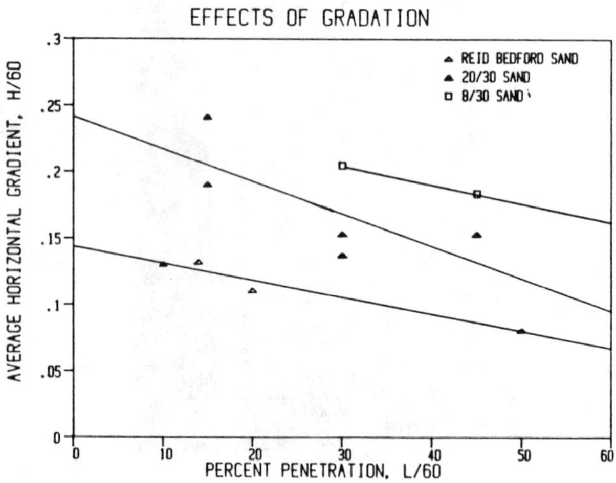

Figure 8: Effects of Gradation on Piping Susceptibility

TABLE 8
SUMMARY OF PIPING TEST RESULTS

Test No. (1)	Sand [(a)] (2)	Pipe Diam (cm) (3)	Penetration % (4)	Ave. Gradient [(b)] (5)	Final Gradient (6)	Density [(a)] (pcf)
1	R.B.	0.635	10	0.130	0.200	106.0
2	R.B.	0.635	10	0.131	0.156	104.8
3	R.B.	0.635	14	0.133	0.138	-
4	R.B.	0.635	20	0.111	0.177	106.2
5	R.B.	3.05	20	0.150	0.112	-
6	R.B.	3.05	50	0.098	0.100	107.4
7	R.B.	3.05	20	0.146	0.158	108.1
8	R.B.	3.05	20	0.142	0.142	-
9	R.B.	0.635	50	0.081	0.127	107.5
10	R.B.	1.27	20	0.133	0.135	92.6
11	20/30	0.635	15	0.19	-	-
12	20/30	0.635	15	0.242	0.298	100.5
13	20/30	0.635	30	0.138	0.383	101.8
14	20/30	0.635	45	0.154	0.166	102.6
15	20/30	0.635	30	0.154	0.212	-
16	WG	0.635	30	N/P	-	110.9
17	WG	0.635	50	N/P	-	112.3
18	8/30	0.635	45	0.183	0.642	106.1
19	8/30	0.635	30	0.204	0.379	106.1
20	Gap I	0.635	15	S/H	-	110.1
21	Gap I	0.635	30	S/H	-	110.4
22	Gap II	0.635	15	0.358 (S/H)	1.03	112.0
23	Gap II	0.635	30	0.958 (S/H)	0.957	113.0
24	Gap II	0.635	45	0.958 (S/H)	0.958	114.0

(a) R.B. = Reid Bedford, WG = well graded
(b) N/P = No Piping, S/H = Self Healing
(c) 1 pcf = $0.16 kN/m^3$

Table 9
CHARACTERISTIC GRAIN DIAMETERS
OF EACH SOIL USED IN PIPING TESTS

	Pipeable			Non Pipeable		
Grain Diam (1)	Reid Bedford (2)	20/30 (3)	8/30 (4)	W.G. (5)	gap I (6)	gap II (7)
D_{10} (mm)	0.14	0.63	0.8	0.24	0.16	0.28
D_{15}	0.17	0.66	0.9	0.35	0.17	0.31
D_{30}	0.19	0.78	1.4	0.92	0.21	0.35
D_{50}	0.20	0.93	1.6	1.42	0.50	0.60
D_{60}	0.21	1.00	1.65	1.60	0.90	1.70
Cu	1.5	1.6	2.1	6.7	5.6	6.1
Cc	1.23	0.97	1.48	2.20	0.31	0.26

$Cu = D_{60}/D_{10}$

$Cc = (D_{30})^2/(D_{10})(D_{60})$

while the uniform sands all piped. Other than C_u and/or C_c, no single granulation characteristic distinguishes piping susceptibility. Although some instability of the finer sizes was observed to move in the gap-graded sands, self-healing occurred ceasing erosion. NHSRI's criteria presented in Table 4 for internal stability are applicable for these well-graded and gap-graded soils. Considering the well-graded sand has $D_3 \approx 0.15$, and $D_o \approx 0.11$ [$D_o = 0.63\ n\ D_{20} = 0.63\ (0.33)(0.55) = 0.11$], the stability observed is consistent with NHSRI's criteria. In the case of the gap-graded sands, for Gap I, $P \approx 42\%$ and for Gap II, $P \approx 47\%$, both of which are $> 35\%$, and thus verify NHSRI's criteria for stability. Although our data are limited, the NHSRI criteria appear valid for evaluating the internal piping susceptibility of sands with $C_u > 5$, while all uniformly graded sands are piping susceptible.

FILTER TESTS

The Reid Bedford sand was used as base soil for the filter tests. Three filter gradations were tested. The grain size distribution curves of these three filter materials and the of Reid Bedford sand are shown in Figure 4. Some characteristic grain diameters of these filters are shown in Table 10. A summary of the filter tests is given in Table 11.

From the test results, it was found that a filter with $D_{15}/d_{85} \leq 8$ would protect the cohesionless base soil properly; where D_{15} is the 15% size of the filter and d_{85} is the 85% size of the base soil. Therefore, a filter design criterion for cohesionless soil can be established as $D_{15}/d_{85} \leq 4$, with a safety factor of two. This conclusion coincides with the criteria proposed by the SCS (see Table 2 soil group 3). Also, the filter design criteria of the U. S. Army Corps of Engineers shown in Table 1, $D_{15}/d_{85} < 5$ (considering piping only), are deemed reasonable and practical.

A comparison was also made between the filter test results and the filter design criteria proposed by NHSRI (see Table 4). If a filter with $D_{20}/d_{70} \leq 10.9$, the filter would protect the base soil properly (see Table 11). Therefore, the criterion for uniform cohesionless soil, $D_{20}/d_{70} \leq 7$, has a safety factor about 1.5 to 2.0.

Filter thicknesses ranging from 1-1/4 to 3/4 in. (31.8 to 19.0 mm) performed satisfactorily in preventing piping and are far thinner than any practicable field filter. However, in the case of filters I and III, as the pipe approached from the downstream side and the filter should have functioned as a "crackstopper", the filters were breached. As the pipe arrived at the downstream side of the filter, its progress was halted and a scouring hole developed. This hole, provided the head was high, would continue to enlarge until support of the filter was removed and the filter collapsed into the hole via a miniature landslide leading to breaching of the filter.

TABLE 10
CHARACTERISTIC GRAIN DIAMETERS OF FILTERS AND BASE SOIL

Grain Diam (1)	Filter I (2)	Filter II (3)	Filter III (4)	Reid Bedford sand (5)
D_{15}	0.85	2.4	3.6	0.17
D_{20}	1.0	2.5	3.8	-
D_{50}	1.55	3.4	4.3	0.20
D_{70}	-	-	-	0.23
D_{85}	-	-	-	0.30

unit: mm

TABLE 11
SUMMARY OF FILTER TESTS

Test No. (1)	Material (2)	D_{15}/d_{85} (3)	D_{20}/d_{70} (4)	Result (5)
16	Filter I	2.83	4.3	Work
17	Filter I	2.83	4.3	Work
18	Filter II	8.0	10.9	Work
19	Filter III	12.0	16.5	Fail

D_{15}, D_{20} = the 15% and the 20% size of the filter
d_{70}, d_{85} = the 70% and the 85% size of the base soil
Base soil: Reid Bedford sand

CONCLUSIONS

1. Piping of sands can be induced and studied visually in the laboratory.
2. Because the horizontal flow velocities acting in sands are insufficient to cause erosion, upward vertical gradients working in combination are required for piping. Therefore, piping requires a discontinuity as an initiator to focus flows and gradients. The greater the pipe penetration, and smaller the discontinuity, the lower the average gradient required to initiate piping.
3. Uniform sands are more susceptible to piping than well-graded or gap-graded sands.
4. In uniform sands, the finer grain sizes are more susceptible to piping.
5. Behavior of the pipe, once initiated in the flume, was contrary to the expectation that the pipe progression should accelerate due to a greater head acting at the tip. Meandering of the pipe led to decreased velocities and particle deposition caused cessation of piping [self-healing] unless the gradient was increased.
6. The SCS filter criterion of $D_{15}/d_{85} \leq 4$ with a factor of safety of two was verified. The COE criterion $D_{15}/d_{85} \leq 5$ is also viable. The NHSRI criterion $D_{20}/d_{70} \leq 7$ has a safety factor of about 1.5 to 2.0.
7. The NHSRI criteria for internal stability can be used to assess the piping susceptibility of well-graded and gap-graded sands.

ACKNOWLEDGMENTS

The substantial contribution of Dr. John H. Schmertmann, Schmertmann and Crapps, Gainesville, Fla., for conceiving and providing valuable discussions to the project are greatly appreciated. The sponsorship of Florida Power and Light and U.S. Bureau of Reclamation for the project is acknowledged.

REFERENCES

Bellport, B. P. (1967), "Bureau of Reclamation Experience in Stablizing Embankment of Fontenelle Earth Dam, 9th Int. Cong. on Large Dams, Istanbul, Turkey pp. 67.

Graf, W. H. (1971), Hydraulics of Sediment Transport, McGraw-Hill, New York.

Liu, Hongmei, and Zhang, Jingmin, "Piping of Sand-Gravel Soils", Northwest Hydrotechnical Science Research Institute, China, 1982.

Liu, J., "Piping Properties of Soils and Filter Design", Northwest Hydrotechnical Science Research Institute, China, 1984.

Richart, F. E. and Schmertmann, J. H. (1958), "The Effect of Seepage on the Stability of Sea Walls," Proc. 6th Conf. on Coastal Engr., Florida Engr. & Industrial Expt. Sta. (EIES), UF, Gainesville, Fla.

Schmertmann, J. H., "A Quantitative Piping Theory and Its Relation to Conditions at the FP&L Dike Breach", Special Board of Consultants Report, Appendix B, 1980.

Seed, H.B. and Duncan, M. J. (1981) "The Teton Dam Failure --- A Retrospective Review," 10th ICSMFE, Stockholm, Vol. 4 pp 214-238.

Sherard, J. L., Dunnigan, L. P., and Talbot, J. R., "Basic Properties of Sand and Gravel Filters", Journal of the Geotechnical Engineering Division, ASCE, June, 1984, pp. 684-700.

Sherard, J. L., and Dunnigan, L. P., "Filters and Leakage Control in Embankment Dams", Proc. ASCE Symp. on Seepage & Leakage from Dams and Impoundments, Denver, CO., May, 1985.

SURFACE EROSION OF COMPACTED PURE CLAYS

By

Alaeddin Shaikh[1], James F. Ruff[2] and
Steven R. Abt[2], Members ASCE

ABSTRACT: Erosion rates of compacted commercially available clays: Na-montmorillonite, Ca-montmorillonite, and kaolinite were measured under a range of tractive stresses from surface flow in a rectangular flume. Samples were compacted under 700 Pa into sample containers with a surface area of 160 cm^2 and were subjected to flowing water in the flume. The compaction water content was near the optimum water content determined from the standard Proctor test. The tractive stress exerted on the bottom of the flume was determined by measuring the velocity distribution of the flow and using the Prandtl-von Karman equation. The tractive stress ranged from 1.67 to 12.9 Pa.

The erosion rate of Ca-montmorillonite was about two orders of magnitude higher than that of Na-montmorillonite and one order of magnitude higher than that of kaolinite. The slaking phenomenon was the cause of relatively high erosion rate of Ca-montmorillonite. Na-montmorillonite eroded by colloidal dispersion, that is, by detachment and removal of particles or particle groups. The erosion rate of kaolinite was influenced by the fabric of the clay particles at the sample surface and by the mixing procedure of dry kaolinite with water.

INTRODUCTION

Surface erosion by rainfall and flowing water can severaly damage earth structures such as earth dams and highway embankments. Surface erosion can occur from overtopping or from sheet or concentrated flow over the embankments. The earth structures are usually constructed from compacted cohesive soils. Therefore, a knowledge of erosion rate of such soils is necessary for design, construction and maintenance of the earth structure. This paper presents the results of studies on erosional behavior of three commerically available clays: Na-montmorillonite, Ca-montmorillonite, and kaolinite. These clays were compacted at near the optimum water content determined by the standard Proctor test and their erosion rates were measured under a unidirectional flow.

---

1 Research Associate, Civil Engineering Department, Colorado State University, Fort Collins, Colorado.

2 Associate Professors of Civil Engineering, Colorado State University, Fort Collins, Colorado.

The results of the erosion studies on relatively pure clays may not represent the erosional behavior of the more complex cohesive soils. Cohesive soils are usually composed of various clay minerals and non-clay material. However, to determine the factors that play the key role in erosional behavior of cohesive soils, studies should begin with the simplest possible conditions. Such a procedure was employed in this study. The influence of parameters such as clay content, water content, and pore water chemistry on erosion rate of compacted pure clays is reported by Shaikh (1986). The influence of clay content, water content, and temperature on erosion rate of compacted soil mixtures containing Na-montmorillonite and Ca-montmorillonite is reported by Grissinger (1966).

Clay Material - Three commerically available clays, two montmorillonites and one kaolinite, were selected for this study. The two montmorillonites are a sodium montmorillonite (No. 200 Volclay bentonite) and a calcium montmorillonite (Panther Creek southern bentonite); both were obtained from the American Colloid Company. An X-ray diffraction analysis showed that both bentonites are composed of well crystallized montmorillonite minerals. The kaolinite was obtained from the Georgia Kaolin Company. The basis for selection of these particular minerals was that (a) they are from the two extreme groups of clay minerals from the viewpoint of particle size and activity, and (b) the proportion of sodium and calcium cations plays a very important role in the behavior of clays. The commercially available clays also have the advantage of purity and uniformity. The physical properties of these minerals are presented in Table 1.

TABLE 1. Physical Properties Of Commerically Available Clay Minerals

Property (1)	Na-Montmorillonite (2)	Ca-Montmorillonite (3)	Kaolinite (4)
Median Size (μm)	< 0.1	4	1.1
Specific Gravity	2.70	2.77	2.60
Liquid Limit	522	90	41
Plastic Limit	41	50	24
Plasticity Index	481	40	17
Optimum Water Content (%)	34	47	24
Optimum Dry Density (g/cm^3)	1.20	1.07	1.57

Note: The optimum water content and optimum dry density were determined from the standard Proctor test.

Eroding Water - Tap water was used as the eroding water and each test started with clean water. The quantity of dissolved salts in the eroding water was less than 2.2 meq/ℓ and the pH of the water was 7.7. The temperature of the eroding water was maintained at 18° ± 1° C by adding hot or cold water during a test.

Sample Preparation - Soil samples were compacted using the following procedures:

1. The optimum water contents for the samples were determined from the standard Proctor test.

2. The soil samples were mixed with tap water at near the optimum water content.

3. For the first sample, an excess amount of wet soil was pressed into the sample container at 700 KPa pressure using a hydraulic press and a specially designed rectangular mold and piston. The surface of the sample was then trimmed until the surface was flush with the top edges of the sample container. The wet weight of the remaining soil in the sample container was considered as the "optimum wet weight." A sample containing material weight equal to the "optimum wet weight" was pressed into the sample container until it produced a soil sample whose surface was flush with the edges of the container.

4. The rest of the samples were prepared by pressing the "optimum wet weight" in the sample container until the surface of the sample was flush with the top edges of the container.

5. Compacted samples were stored in plastic bags for 24 hours before the erosion test.

Testing Apparatus and Procedures - The erosion tests were performed using a rectangular Plexiglas flume 15.5 cm wide, 11 cm deep, and 250 cm long with an adjustable slope. A schematic drawing of the flume and sample containers are shown in Figure 1. The sample containers were 15.2 cm long, 10.5 cm wide, and 2.25 cm deep. The flume floor was modified so that the surface of the sample was flush with the bottom of the flume. Water was recirculated in the system. The depth of flow ranged from 0.8 to 2.1 cm. The flow rate was measured by a precalibrated Venturi meter and uniformity of the flow depth was manipulated by an upstream control. The velocity profiles along the flume at sections shown in Figure 1 were measured using a 0.3 cm O.D. pitot tube connected to a pressure transducer and readout system.

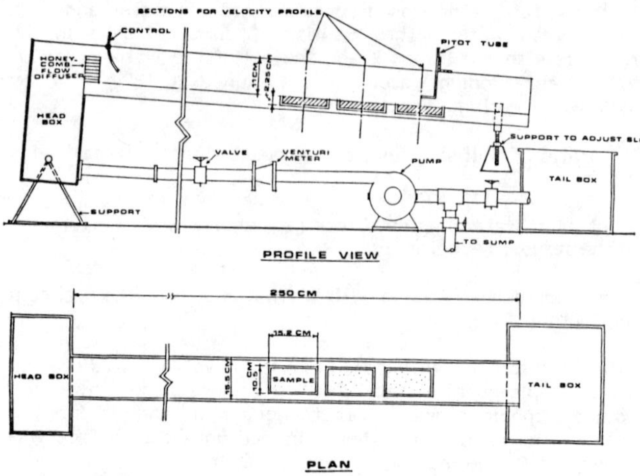

Figure 1. Schematic Drawing Of The Flume and Sample Containers

The tractive stress, τ, exerted on the flume bottom was estimated by measuring the velocity profiles and using the Prandtl-von Karman equation for smooth channels (Chow 1959).

$$\frac{V}{V_*} = 5.5 + 5.75 \log \left(\frac{V_* y}{\nu}\right) \qquad (1)$$

where $V_* = \tau/\rho$ is shear velocity, V is velocity at the depth y in the turbulent region, ρ is density of water, and ν is kinematic viscosity of water. The tractive stresses were determined for five locations across and along the flume and then averaged. The average of tractive stress values obtained by this method was in close agreement with the value of τ obtained using deBoys equation (Chow 1959),

$$\tau = \gamma R \operatorname{Sin} \alpha \qquad (2)$$

where γ is unit weight of water, R is hydraulic radius, and α is angle between the flume and horizontal surface. For small flow depth and high slopes where velocity profile measurement was not possible, the tractive stress was estimated by Equation 2.

For each soil sample a total of nine tests were conducted at three flume slopes and three time durations. To reduce the testing time, the flume was designed so that up to three samples could be tested during the same experiment. When the erosion rate of the samples was relatively low, as in the case of Na-montmorillonite samples, three samples were placed in the flume for each run. For Ca-montmorillonite samples and kaolinite samples, where the erosion rate was relatively high, only one sample per run was used. In the latter case, the rest of the sample locations in the flume were filled with dummy samples made of Plexiglas.

The amount of erosion (weight loss) was defined as the difference between the dry weights of the samples before and after testing. The erosion rate is defined as the slope of the line relating weight loss versus time. This value is determined by using a least square fit of the data and by imposing the restriction that the lines must pass through the origin, i.e., at time zero weight loss must be zero. The value of the tractive stress for zero erosion rate was defined as the critical tractive stress.

EXPERIMENTAL RESULTS AND DISCUSSION

In addition to erosion tests, the reaction of clay samples to still water was observed. In this section the reaction of clay samples to still water will be discussed. Then the erosion test results will be presented and discussed.

Behavior of Clays in Still Water - One sample of each clay, identical to the samples prepared for the erosion tests, at optimum water content, was immersed in water for 10 minutes and any reaction was noted. The Na-montmorillonite sample did not show appreciable swell, although dispersion of the surface particles was noticed. The surface of the sample became sticky, as was the case in the erosion tests. The Ca-montmorillonite sample, however, swelled and its thickness increased by about 50%. The kaolinite sample also swelled and its thickness increased by about 25%. Upward flow of entrapped air in a few locations carried water laden kaolinite upward resembling an eruption. A similar but less intense reaction was observed with the Ca-montmorillonite. The depth of penetraton of water into the Na-montmorillonite sample was less than a millimeter, while for the Ca-montmorillonite and kaolinite samples, water completely penetrated the samples.

The behavior of the Ca-montmorillonite and kaolinite in still water is characterizied by slaking. Slaking is the breakdown of soil aggregates upon immersion in water. When an unsaturated soil specimen is immersed in water, the water will penetrate into the speciment under the driving force of the matrix and osmotic potential. The air in the voids will be trapped between menisci of the entering water, and air pressure will build up as the water fills the voids. The entrapped air will then exert an enormous pressure causing a rapid increase in void size and disintegration of the soil (Terzaghi and Peck 1967, Sowers 1979, Winterkorn 1973).

General Observations for Erosion Tests - The mechanism of erosion was observed to vary for different clays tested. The Na-montmorillonite eroded particle by particle. At the beginning of each test the recirculating water was clear and gradually became cloudy. The Ca-montmorillonite eroded in large masses, while the erosion of kaolinite took place by detachment and removal of large soil aggregates with sizes up to 5 mm. The pattern of erosion for all the Na-montmorillonite samples with all clay fractions was similar to that shown in Figure 2. The three samples shown in Figure 2 represent various erosion times. Except for the dips at the upstream and downstream ends, the remainder of the sample surface remained flat and smooth after erosion. The form of erosion at the upstream and downstream ends of the samples is believed to be a result of eddies that have developed from changes in roughness of the flume bed when the bed changed from the Plexiglas to a soil sample surface. The erosion time of the Na-montmorillonite samples ranged from 30 to 120 minutes and the erosion depth varied from 1 to 6 mm. The depth of penetration of water into the samples during the erosion tests was less than one millimeter.

Figure 2. Erosion pattern of Na-montmorillonite samples. Samples are for three different times of erosion, increasing in duration from top to bottom.

No specific erosion pattern was observed for Ca-montmorillonite as it had a high erosion rate, two orders of magnitude higher than that of Na-montmorillonite. An erosion pattern of Ca-montmorillonite is shown in Figure 3. Most of these samples eroded in large masses of soil aggregates. The erosion time of the Ca-montmorillonite samples ranged from 45 seconds to 4 minutes. After the erosion tests, the remaining clays in the sample containers were completely wet and soft.

The mechanism of erosion of the compacted kaolinite was quite different from that of the montmorillonites. The erosion pattern of the kaolinite samples is shown in Figure 4. Large aggregates of kaolinite with sizes up to 5 mm were observed to detach from the bed and then were transported away. These soil aggregates formed during sample preparation when dry kaolinite was mixed with water and did not break apart when the

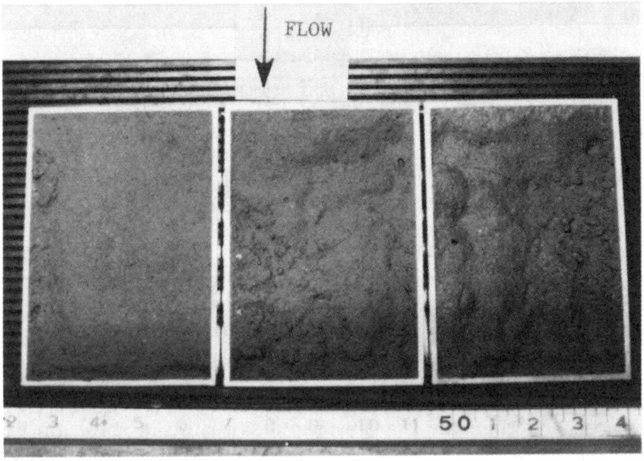

Figure 3. Erosion pattern of Ca-montmorillonite samples. Samples are for three different times of erosion, increasing in duration from left to right.

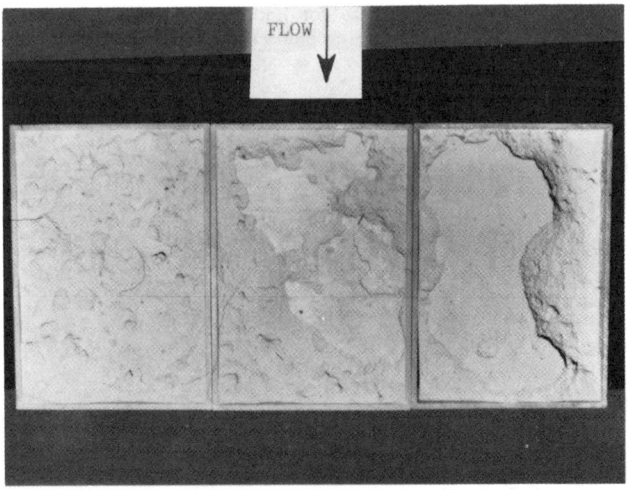

Figure 4. Erosion pattern of kaolinite samples compacted at 24.1% water content. The surfaces of the two samples on the right have been trimmed. The erosion time for the samples from left to right was 30, 60, and 90 min, respectively.

soil was compacted, even though they may have deformed. Formation of kaolinite aggregates during sample preparation is also reported by Hodek and Lovell (1979). Hence, when kaolinite samples were exposed to running water, the erosion occurred from detachment and movement of these aggregates. It was observed that the structural arrangement of clay particles on the sample surface was very important in the erosion resistance of kaolinite. The samples with trimmed surfaces showed a much higher resistance to erosion than those that were not trimmed. But after the surface layer of the trimmed samples was eroded, the remainder of the sample showed a very low resistance to erosion as shown in Figure 4. The erosion time for the kaolinite samples which were trimmed ranged from 60 to 120 minutes. But for the samples which were not trimmed, the erosion time ranged from 5 to 60 minutes. After the erosion test, the remaining kaolinite was completely wet and very soft.

Erosion Test Results - A typical relationship between weight loss per unit surface area and time for the clay samples tested in shown in Figure 5. The magnitude of erosion rate for given tractive stresses also are shown in Figure 5 for Na-montmorillonite sample. Most of the data for Na-montmorillonite and Ca-montmorillonite showed a linear relationship between weight loss and erosion time. The erosion rate values shown in Figure 5 are determined by using a least square fit of the data and by imposing the restriction that the lines must pass through the origin. The weight loss versus erosion time for kaolinite samples did not follow a specific pattern. The erosion rate for kaolinite samples, therefore, was determined as the average of the erosion rate (weight loss/time) of three samples tested at the same conditions.

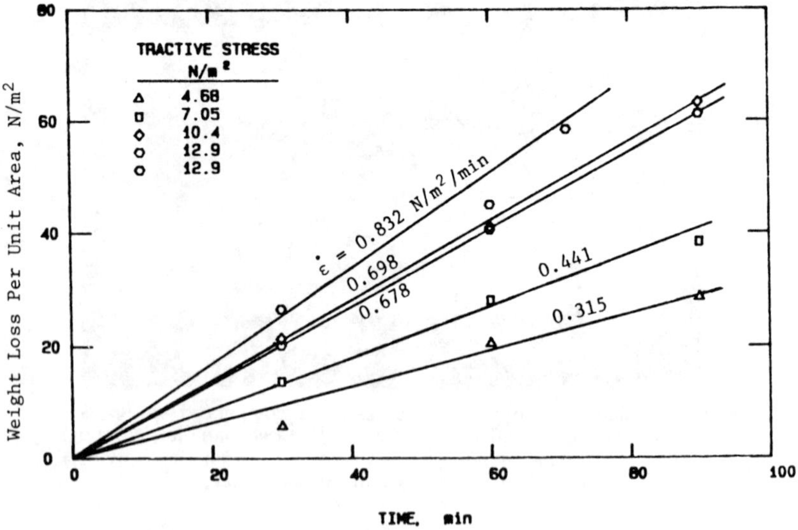

Figure 5. Weight Loss Per Unit Area Versus Time for Na-Montmorillonite Compacted at 34.5% Water Content.

The relationship between erosion rate ($N/m^2/min$) and tractive stress (N/m^2) for the three clays tested is shown in Figure 6. For Na-montmorillonite the erosion rate varied linearly with tractive stress as shown in Figure 6. For Ca-montmorillonite and kaolinite samples the data are scattered and do not follow a specific pattern. However, for comparison purposes, dashed lines are drawn through the data points to indicate the average variation of erosion rates versus tractive stress. The slope of the lines of erosion rate versus tractive stress is defined as the erosion rate coefficient which is in units of min^{-1}.

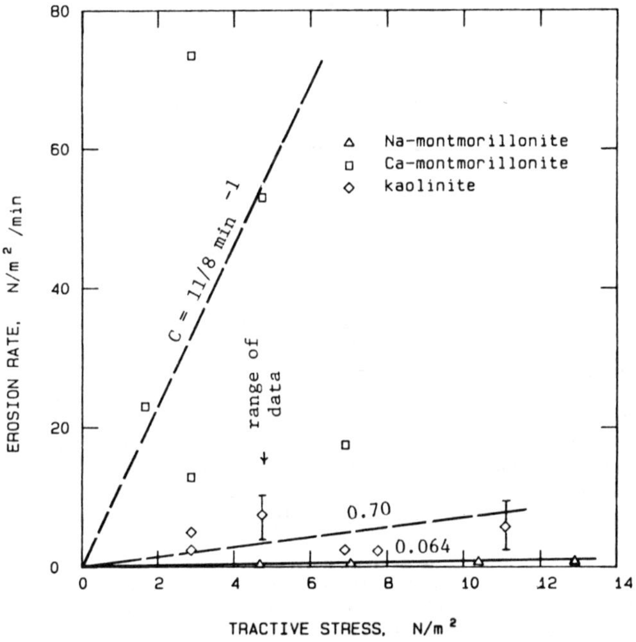

Figure 6. Erosion Rate Versus Tractive Stress

The values of erosion rate coefficient and chemical properties of the clays tested are presented in Table 3. The chemical properites of clays are determined from a saturation extract. In Table 3 the shear strength values of the as compacted samples that were determined by a Torvane apparatus are also presented. Torvane shear strength was determined using compacted samples identical to samples prepared for erosion tests.

TABLE 3 Chemical Properties And Erosion Rate Coefficient Of The Clays Tested.

Clay (1)	SAR[a] $(meq/\ell)^{1/2}$ (2)	TDS[b] (meq/ℓ) (3)	Percent[c] Saturation (4)	pH (5)	Erosion Rate Coeff. min^{-1} (6)	Torvane Shear Strength KPa (7)
Na-montmorillonite	19.8	20.5	755.6	9.8	0.064	2100
Ca-montmorillonite	0.4	7.8	96.6	8.2	11.8	2048
Kaolinte	0.5	1.7	98.3	5.5	0.7	1104

a SAR = Sodium adsorption ratio = $Na/\sqrt{0.5\ (Ca+Mg)}$
b TDS = Total dissolved salts = Ca + Mg + Na + K
c Percent Saturation = water content of the paste used for saturation extract

Ca, Mg, Na, K = Calcium, Magnesium, Sodium, and Potassium cations

The erosion rate coefficient of Ca-montmorillonite was about two orders of magnitude higher than that of Na-montmorillonite and one order of magnitude higher than that of kaolinte as shown in Table 3. The results of this study were in agreement with those reported by Grissinger (1966). Grissinger measured the erosion rate of soil mixtures containing Na-montmorillonite and Ca-montmorillonite. His results show that the soil containing Ca-montmorillonite was more erodible than the soil containing kaolinte and Na-montmorillonite. Grissinger used a similar method of compaction to this study and performed his erosion tests in a flume.

The phenonmenon of slaking was belived to be the cause of the high erosion rate of the Ca-montmorillonite in this study. Slaking is the breakdown of soil aggregates upon immersion in water. The low SAR (sodium adsorption ratio) Ca-montmorillonite has a flocculated fabric and therefore is highly permeable. As a result, water easily penetrates into the clay and slaking occurs. Na-montmorillonite has a dispersed fabric with a low permeability and, therefore, water cannot easily penetrate into the soil. Hence, soil does not slake and erosion occurs only by detachment and removal of the clay particles or particle groups at the sample surface. The chemical properties of the saturation extract from the Ca-montmorillonite and kaolinite were nearly the same. Therefore, the difference in the erosional behavior of these two clays could be attributed to their mineralogy. The kaolinite and montmorillonite behaved differently when mixed with water. When dry kaolinite was mixed with water, aggregates of particles formed, while montmorillonite clays (both the sodium and calcium types) mixed uniformly. This mechanism of mixing influenced the fabric of the compacted samples and, consequently, the mechanism of erosion. The size and distribution of the kaolinite aggregates varied for samples prepared at different times. Since the erosion of kaolinite samples occurred by detachment and movement of these aggregates, the erosion rate was dependent on the size and distribution of aggregates and/or the procedure for mixing the dry kaolinite with water. The data scatter for kaolinite is believed to be because of variation of the size and distribution of the aggregates formed during sample preparation.

The Na-montmorillonite and Ca-montmorillonite samples had nearly identical values of shear strength as shown in Table 3, but the coefficient of erosion rate is two orders of magnitude higher for Ca-montmorillonite. These results indicate that there is not a correlaton between erosion resistance and as compacted Torvane shear strength.

CONCLUSIONS

The order of increasing erosion rate of the three clays tested was as follows:

Ca-montmorillonite > kaolinite > Na-montmorillonite

Both Ca-montmorillonite and kaolinite slaked when immersed in still water. Na-montmorillonite, however, showed no immediate reaction to still water. Na-montmorillonite was observed to erode particle by particle during the erosion test, that is, by colloidal dispersion. Ca-montmorillonite eroded in masses of small soil aggregates. It is believed that the slaking phenomenon was the cause of relatively high erosion rate of Ca-montmorillonite. The erosion rate of kaolinite was influenced by the fabric of the clay sample. Kaolinite eroded by detachment and removal of the aggregates formed during sample preparation. The tractive stress for zero erosion rate (critical tractive stress) was zero for the three clays tested.

A relationship was not found between the erosion rate and Torvane shear strength of the clays tested. The Torvane shear strengths of Na-montmorillonite and Ca-montmorillonite were nearly the same, but the erosion rate of Ca-montmorillonite was two orders of magnitude higher than that of Na-montmorillonite.

APPENDIX 1. - References

Chow, V.T. (1959). Open-Channel Hydraulics. McGraw-Hill, New York, pp. 168 and 200.

Grissinger, E. H. (1966). "Resistance of Selected Clay Systems to Erosion by Water." Water Resources Research, Vol. 2, No. 1, pp. 131-138.

Hodek, R. J. and Lovell, C. W. (1979). "Soil Aggregates and Their Influence on Soil Compaction and Swelling." Transportation Research Record 733, National Research Council, pp. 94-99.

Shaikh, A., (1986). "Surface Erosion of Compacted Clays," thesis presented at Colorado State University, Fort Collins, Colorado, in partial fulfillment of the degree of Doctor of Philosophy.

Sowers, G. F. (1979). Introductory Soil Mechanics and Foundations: Geotechnical Engineering, 4th edition, Macmillan Publishing Co., Inc., New York.

Terzaghi, K. and Peck, R. B. (1967). Soil Mechanics in Engineering Practice, 2nd Edition, John Wiley and Sons, Inc., New York, pp. 146.

Winterkorn, H. F. (1973). "Experimental Study of the Attack of Water on Dry Cohesive Soil Systems," Soil Erosion: Causes and Mechanisms; Prevention and Control, Special Report 135, Highway Research Board, National Research Council, pp. 1-7.

Piping Failure of a Non-Dispersive Clay Dam

By Dan A. Brown[1] and R. Kenneth Derick[2]

ABSTRACT: A low earthen embankment with a 3-ft thick clay lined impoundment was constructed to contain industrial wastewater at a site in southwestern Louisiana. In spite of the fact that the dam and the liner were constructed of well-compacted non-dispersive clay, a piping failure occurred within a few months after filling the reservoir. Investigation into the failure revealed that an old slough beneath the embankment had been filled with soil which was much more compressible than the surrounding natural stiff clays. Differential settlement beneath the embankment resulted in horizontal cracking and piping through the stiff embankment soils above the old slough. This case history demonstrates that piping can occur even in low dams built of non-dispersive clays where foundation conditions are conducive to severe local differential settlement.

INTRODUCTION

This paper provides a case history of a piping failure of a low dam constructed of well-compacted and non-dispersive clay. The dam was constructed across an old slough which had previously been filled with soft, compressible material. The differential settlement which occurred above the old slough resulted in seepage erosion, possibly produced by hydraulic fracturing. This case history illustrates the importance of relatively uniform foundation support, even in relatively low dams.

SITE CONDITIONS

The site consists of stiff, overconsolidated silty clays (CL) of Pleistocene Age, known locally as the Bentley Formation. Liquid limits range from 40 to 50 with plasticity index values in the range of 20 to 30. Natural water contents were at or slightly above the plastic limit, typically in the range of 18 to 22%.

The facility serves as an industrial wastewater pond, and will eventually be filled with papermill ash. The area near the failure is shown in plan on Figure 1, and in cross-section on Figure 2. The embankment was constructed of materials excavated from the interior

---

[1] Assistant Professor, Auburn University, Auburn, Alabama 36849, formerly Chief Engineer, Soil Testing Engineers, Inc., P.O. Box 80379, Baton Rouge, Louisiana 70898.

[2] Branch Manager, Soil Testing Engineers, Inc., 4313 Downtowner Loop, North, Mobile, Alabama 36609.

of the pond. The silty clay soils were compacted in 8 inch horizontal lifts to at least 95% Standard Proctor density, with each lift scarified before placing the next lift. Liner materials were placed in 8 inch loose lifts parallel to the surface and compacted to similar density requirements. Fixed-wall permeability tests were performed on drive-tube samples taken from the in-place liner to confirm that the hydraulic conductivity was less than 1.0x10(-7) cm/sec.

Other water retention facilities in this area have been constructed to heights at least twice as great as the ash pond levee, with no significant seepage problems. Erosion due to rainfall runoff has been a bit troublesome, requiring slope maintenance every few years.

In the area of the piping failure, a ditch had been present some years prior to the ash pond levee construction. This ditch was about 10 feet deep and had been filled with loose, uncompacted gray silts containing significant organic matter. According to records obtained after the failure occurred, the new ash pond levee crossed the filled-in ditch approximately as shown on Figure 1.

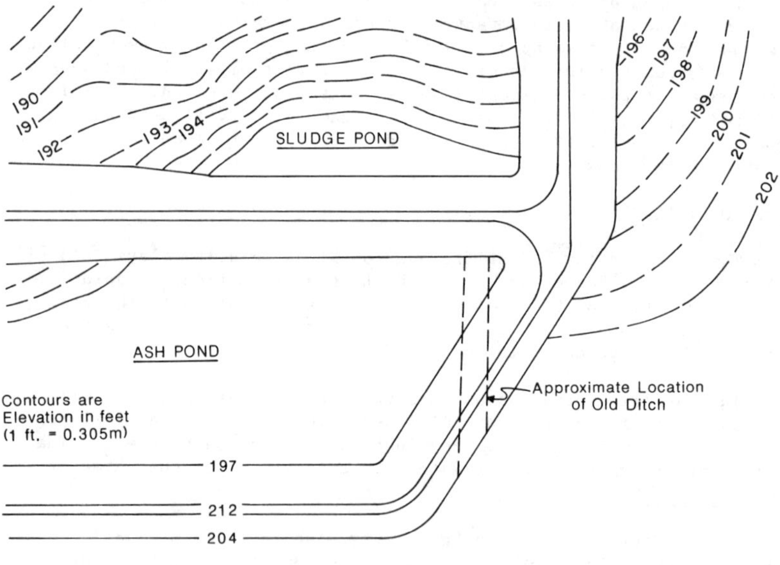

Figure 1. Plan View of Site.

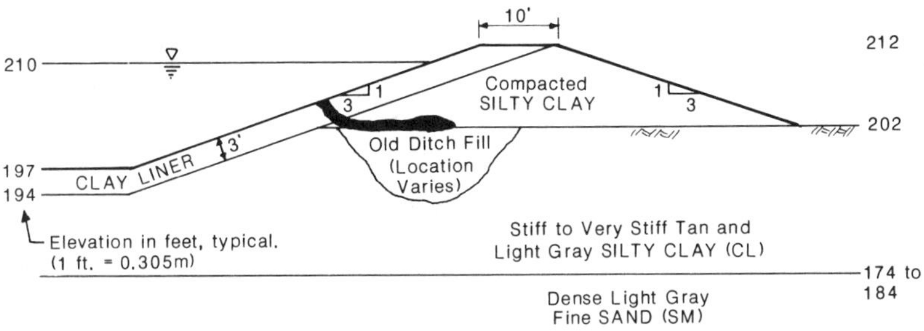

Figure 2. Cross-Section of Levee.

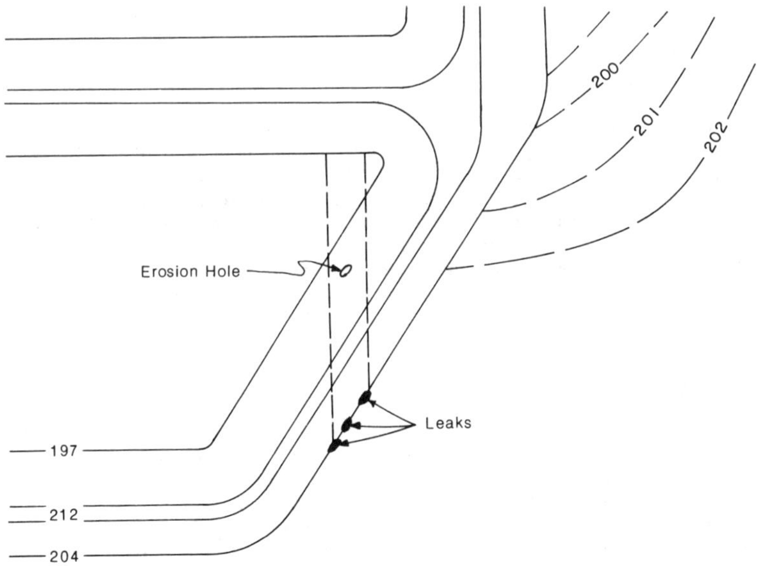

Figure 3. Location of Observed Seepage.

DESCRIPTION OF FAILURE

The ash pond was not filled until approximately one year after levee construction was completed. Approximately two to three months after filling, engineers at the mill noticed an unusual amount of seepage at a particular spot on the outside of the levee. Upon observing an overnight increase in the quantity of seepage, the mill maintenance workers attempted to stop the leak by placing a mound of earth over the outflow. Within one day, two new leaks occurred about 60 feet south of the first. Small levees were then constructed to contain the outflow from all three locations, and the ash pond level was lowered. After lowering the pond level, a single erosion hole was found on the interior of the levee. The location of the erosion hole and the exterior leaks are shown on Figure 3, and a photograph of the interior erosion hole is shown on Figure 4. There was no evidence of cracking on the face of the levee which could have directed a concentrated flow at this particular location.

As the levee was excavated for reconstruction, a seepage erosion channel could be traced along the location of the filled-in ditch. The erosion appeared as a relatively flat and wide channel at the interface between the embankment and the top of the ditch fill (see Figure 5). Substantial pockets of water were encountered along this interface during the excavation. The levee was repaired by excavating the entire area above and through the filled in ditch, and replacing with compacted clays as per the original construction.

Because of the problem with seepage erosion, undisturbed samples of the compacted silty clay fill from the ash pond and other levees in the area were tested for dispersibility. Pinhole tests were conducted in accordance with guidelines of Sherard, Dunnigan, Decker and Steel (1976) and crumb tests in distilled water were performed following the procedure described by Sherard, Dunnigan and Decker (1976). All of the tests performed indicated the soil to be non-dispersive. These results tend to compare well with local experience prior to this failure, in which seepage erosion has not been a problem.

EXPLANATION

It appears that the seepage erosion problem at this site can be attributed to the differential settlement of the foundation beneath the dam. The compressibility of the loose fill in the old ditch relative to that of the natural stiff clay appears to have results in arching and a zone of very low vertical stress in the area immediately above the ditch. It is quite likely that this condition led to formation of the initial leak by hydraulic fracturing, as outlined by Sherard, Decker and Ryker (1972). As stated in that article:

Figure 4. Erosion Hole on Interior of Levee.

> "While damage or failure by piping from initial leaks starting by hydraulic fracturing is far more likely to occur in homogeneous dams of dispersive clay, it is believed that under extreme conditions similar failures have occurred in homogeneous dams of ordinary non-dispersive clay."

This case history appears to represent such an extreme condition. It is also likely that a cavalier attitude toward foundation treatment was encouraged by both the absence of any history of seepage erosion problems in homogeneous clay dams in this area and by the use of 3 foot (1m) thick clay liners on the interior of all wastewater impoundments.

Although the piping mechanism at the interface between the embankment and the soft ditch-fill appears to be reasonably well understood, the mechanisms producing the piping through the clay-lined upstream face are not obvious. Certainly there were zones of low total stress in the area where differential settlement occurred. Hydraulic fracturing could initiate in small cracks from differential settlement, as well as surface cracks from drying prior to pond filling.

Figure 5. Internal Seepage Channel Above Compressible Ditch Fill.

LESSONS LEARNED

The case history presented in this paper should serve to remind designers of several important considerations in design of wastewater impoundments:

* Even small homogeneous and well-constructed dams of nondispersive clay can be subject to seepage erosion problems if the potential for differential settlements is significant.

* Reliance solely upon an internal liner as an impermeable barrier to flow is imprudent; a well-constructed liner is only one part of the total containment structure.

* Where clay materials are to be used as the primary flow barrier, the surface of these clays should be protected from drying.

ACKNOWLEDGEMENTS

The writers gratefully acknowledge the contribution of the management of Soil Testing Engineers, Inc. for providing release from other duties to prepare this paper and to consider this case history in detail.

REFERENCES

Sherard, J.L., Ryker, N.L., and Decker, R.S., "Hydraulic Fracturing in Low Dams of Dispersive Clay," Proceedings, Specialty Conference on the Performance of Earth and Earth-Supported Structures, ASCE, Vol. 1, 1972, pp. 653-689.

Sherard, J.L., Dunnigan, L.P., Decker, R.S., and Steel, E.F., "The Pinhole Test for Identifying Dispersive Soils," Journal of the Geotechnical Engineering Division, ASCE, Vol. 102, No. GT1, Jan., 1976, pp. 69-85.

Sherard, J.L., Dunnigan, L.P., and Decker, R.S., "Identification and Nature of Dispersive Soils," Journal of the Geotechnical Engineering Division, ASCE, Vol. 102, No. GT4, Apr., 1976, pp. 287-301.

DISPERSIVE SOIL AND ITS INFLUENCE ON THE DESIGN OF REACH 1A OF THE TUCSON AQUEDUCT - CENTRAL ARIZONA PROJECT

Richard L. McNearny*

INTRODUCTION

The Central Arizona Project is a USBR (U.S. Department of the Interior, Bureau of Reclamation) multipurpose water resource development and management project which will provide supplemental Colorado River water, either directly or by exchange, to nearly 1 million acres of irrigated agricultural land in Maricopa, Pinal, and Pima Counties in Arizona and western New Mexico (fig. 1). Approximately 500,000 acre-feet of municipal and industrial water will be provided for the Phoenix and Tucson metropolitan areas. Additional benefits include power generation, flood control, recreation, fish and wildlife conservation, and sediment control. The project is divided into seven administrative and construction divisions: Granite Reef, Regulatory Storage, Salt-Gila, Gila River, Tucson, Indian Distribution, and Colorado River (Project Data, 1981).

Tucson Aqueduct begins at the terminus of the Salt-Gila Aqueduct and will transport water 60 miles to Tucson, Arizona. Reach 1A of Tucson Aqueduct will be approximately 5 miles of concrete-lined canal with a beginning carrying capacity of 2,250 ft^3/s (fig. 2). The canal will carry water year-round. Any unexpected break in service to water users will have to be resolved quickly, as uninterrupted water deliveries to the Tucson area are top priority. This paper discusses only Reach 1A of Tucson Aqueduct, as the problem of dispersive soil was extensive for this reach. Dispersive soils were first encountered on Reach 3 of the Salt-Gila Aqueduct, but the problem was limited.

During preconstruction investigations for Reach 1A of the aqueduct, it was discovered that the basin fill deposits along the proposed canal alignment contained a high percentage of dispersive soil and thus were considered unacceptable for canal and collective dike compacted embankment. The collective dike intercepts and retains cross drainage flow. The impounded water is directed to culverts, which transport the collected water safely underneath the canal. Further, it was anticipated that the canal prism might be excavated into dispersive soils immediately overlying pervious material. This situation would provide a possible outlet for dispersive material to pipe into pervious material underneath with resultant possible collapse of the canal lining through loss of foundation support.

*Civil Engineer, Bureau of Reclamation, Engineering and Research Center, Denver, Colorado.

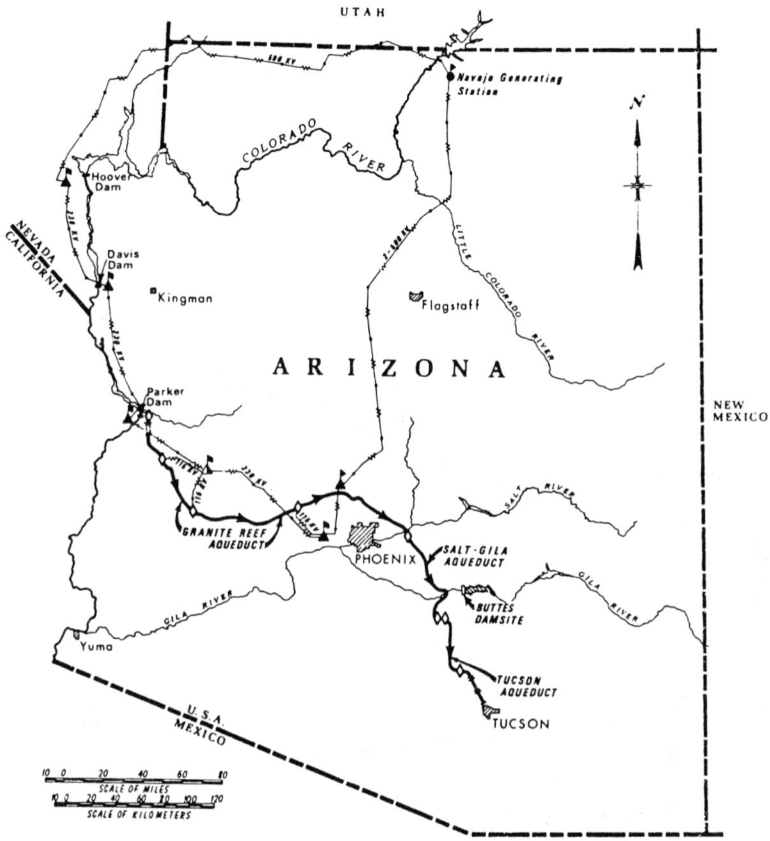

Figure 1. - Central Arizona Project (Project Data, 1981)

Further investigations were initiated to define the extent of the problem along the canal alignment, to locate borrow sources of nondispersive soil (fig. 2), and to formulate an economical design which would permit construction of a canal through dispersive soil.

SITE GEOLOGY

Reach 1A, station 110+00 to station 369+50, of the Tucson Aqueduct is located within the Basin and Range Physiographic Province of Arizona (Wellendorf, 1982). The province is characterized by generally north- to northwest-trending uplifted mountain ranges separated by broad, down-dropped basins. The basins are filled with thousands of feet of alluvial or lacustrine deposits with some interbedded volcanic rocks. The basin fill deposits are thin near the range fronts and interfinger with the alluvial fan deposits which ring the ranges. The deposits grade from coarser to finer grained with increasing distance from the

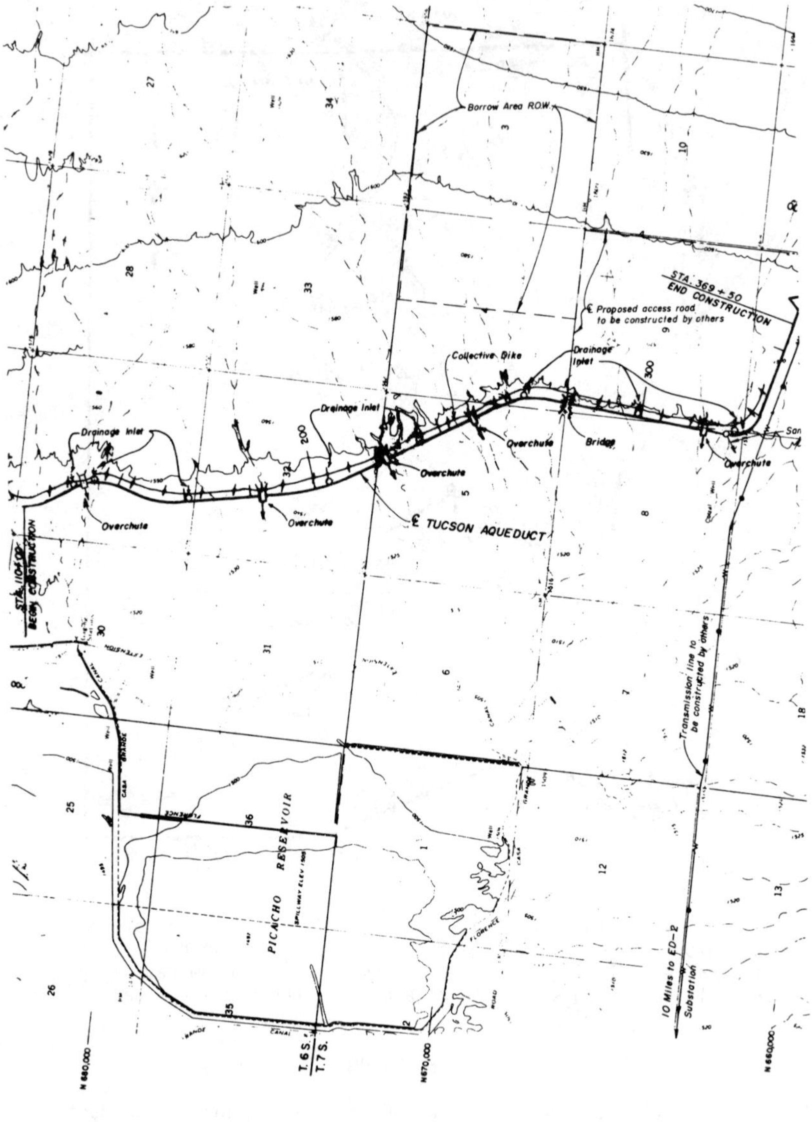

Figure 2. - Tucson Aqueduct, Reach 1A

mountains. The basin fill deposits consist predominantly of clayey sand, sandy clay, and silty sand, with lesser amounts of lean clay and sandy silt and thin lenses of poorly to well graded sand. These sediments were deposited by outer edge alluvial fan processes such as sheet flooding and ephemeral streamflow. Subsidence west of the aqueduct alignment has caused rapid downcutting and headward erosion of streams crossing the basin fill.

Drainage is generally from east to west and crosses the aqueduct alignment at nearly right angles (fig. 2). Most of the drainages are typical arid-region arroyos, carrying water only during and immediately after periods of heavy rain. The drainages generally contain thin accumulations of predominantly clean, coarse sand not extensive enough to be shown on geologic maps.

GEOTECHNICAL INVESTIGATION AND LABORATORY TESTING

Feasibility investigations began in 1965 for this reach of the aqueduct. Seven auger holes and one drill hole were drilled and one test pit was excavated in the vicinity of the alignment.

Preconstruction investigations of Reach 1 were begun in May 1982. Fourteen bucket auger holes were drilled approximately 1,800 feet apart along the alignment of the canal centerline. Two additional auger holes were drilled along the alignment and two auger holes were drilled for the abutments at a bridgesite. The holes were augered using a 24-inch-diameter bucket. Three test pits were excavated in June and July 1982. Two drill holes were drilled at a bridgesite using a Pitcher-type core barrel sampler to obtain undisturbed samples for laboratory testing.

Thirty-four Pitcher-type samples and twenty-six sack samples were shipped to the E&R (Engineering and Research) Center geotechnical laboratory in Denver, Colorado, for testing. Gradation analyses, Atterberg limits, specific gravity, laboratory compaction tests, and dispersive clay tests were performed on seven samples; and gradation analyses, Atterberg limits, and dispersive clay tests were performed on the remaining samples.

All tests, except for the dispersive clay tests, were performed in accordance with procedures outlined in the _Earth Manual_ (1980). "Double hydrometer" tests (Volk, 1937) and "crumb" tests (Emerson, 1967) were performed in accordance with procedures outlined in _Laboratory Procedures for Determining the Dispersibility of Clayey Soils_ (Kinney, 1979). Pinhole tests were performed in accordance with procedures described in _Quantification of Pinhole Test Equipment Hydraulic Characteristics_ (Acciardi, 1982). Pinhole test specimens were compacted to an approximate "target" dry unit weight corresponding to a moisture content 2 percent dry of optimum, as determined by the laboratory compaction test (12,375 ft-lbf/ft^3 of compactive effort). Chemical tests were not performed on pore fluid extracts because the currently accepted chemical test methods (Sherard, et al., 1976) fail to adequately corroborate physical test results (Craft and Acciardi, 1984).

Twenty-one of the samples tested were dispersive, four were intermediately dispersive, and one was nondispersive. Eight samples were not tested since soil with a PI (plasticity index) equal to or less than 7 percent, or which contains 12 percent or less minus 0.005-millimeter-sized particles, is generally considered erodible whether the clay fines are dispersive or not (Kinney, 1979).

Material is normally processed through the No. 4 sieve prior to wetting and compaction of the test specimen for the pinhole test (Kinney, 1979). However, because of the sand present (up to 50 percent by dry weight) in the majority of the test specimens, dislodged coarse sand particles would frequently interfere with the flow of water through the 1-millimeter-diameter hole in the test specimen. Consequently, other sieve sizes were evaluated for preparing test specimens. The most satisfactory results were obtained when the material was screened through a No. 30 sieve, i.e., all the coarse and most of the medium sand particles were removed.

As a result of the investigation and test programs, the material to be excavated from the canal prism between stations 128+60 and 146+00 and stations 214+00 and 369+50 was identified as largely dispersive and unacceptable as construction material for canal and collective dike embankment. The remainder of the material along the alignment was generally identified as silty sand having little or no cohesion and thus susceptible to erosion from running surface water.

Additional extensive field explorations and laboratory tests were conducted by the Arizona Projects Office in Phoenix, Arizona, to locate and delineate borrow materials suitable for construction (Whitaker, 1983). These explorations were conducted both along the proposed canal alignment and within a distance of 3 miles to the east of the alignment. A borrow area was subsequently located 1 to 2 miles east of the alignment at about station 260+00 (fig. 2).

DESIGN

The overall design philosophy for Tucson Aqueduct was to (1) not use dispersive soil as the core material in water-retaining structures, (2) slow the movement of rainfall on the top and downslope side of embankments by use of windrows and ponds, and (3) use erodible materials only where they could be protected from moving water and against piping. Further, there was concern that water leaking through the canal lining could undermine the lining in canal reaches cut into in situ materials by piping dispersive soils into any pervious lenses or layers below the dispersive soil. Some of the drill logs indicated a strong possibility that pervious lenses and layers would be encountered in the arroyos, and further, that the lenses and layers extended upstream into the reservoir areas behind the collective dikes. Sand lenses open to the reservoir could carry water to the concrete lining, causing uplift, with resultant breakage of the lining if the canal was dry.

In an attempt to reduce the amount of material to be wasted, a decision was made that dispersive soil either had to be treated to render it nondispersive or it had to be incorporated into the design.

A limited study was conducted to evaluate the effects of treatment of dispersive soil with lime or fly ash (McNearny, 1985).

Lime-treated specimens from two samples of dispersive soil were prepared at 0.5, 1, 2, and 4 percent hydrated lime by dry weight of soil. The two samples represented the range of plasticity encountered in the soils classified as dispersive. Atterberg limits, "crumb" tests, and pinhole tests were performed on the specimens of lime-treated soil. The PI decreased from 30 percent (untreated) to 3 percent (at 4 percent lime) for one sample and from 6 percent (untreated) to 2 percent (at 4 percent lime) for the second sample. All of the lime-treated specimens from the first sample were nondispersive. For the second sample, the specimen containing 0.5 percent lime was intermediately dispersive; but the specimens containing 1, 2, and 4 percent lime were nondispersive.

Fly ash-treated specimens of the same soil were prepared at fly ash ratios of 1:1, 1:3, 1:7, and 1:15 (dry weight of fly ash to dry weight of soil). Atterberg limits, "crumb" tests, and pinhole tests were performed on the specimens of fly ash-treated soil. The PI decreased from 30 percent (untreated) to 5 percent (1:1, fly ash:soil) for the first sample and from 6 percent (untreated) to nonplastic (1:1, fly ash:soil) for the second sample. All of the fly ash-treated specimens from the first soil sample were nondispersive. The specimens from the second soil sample with fly ash to soil ratios of 1:15 and 1:3 were intermediately dispersive, and the specimens with ratios of 1:7 and 1:1 were nondispersive. Chemical analyses of fly ash specimens indicated that distribution of calcium oxide within the fly ash was uneven, which could explain why the 1:3 specimen was intermediately dispersive, while the 1:7 specimen was nondispersive. Thus, even if a well proportioned mix of fly ash and soil could be obtained in the field, there was no guarantee the amount of calcium oxide needed to properly treat the soil would be available.

With increasing amounts of lime or fly ash added to the soil, the PI of the samples decreased. With decreasing PI, cohesiveness of the soil decreased with a corresponding increase in potential for erosion to occur. Since treating the dispersive soil only increased erosion potential and since there was already too much soil considered erosive, i.e., silty sand, soil treatment was eliminated as an alternative for design. Economic analysis performed later also indicated that soil treatment would have been prohibitively expensive (Straubinger, 1983).

The design of the canal, therefore, had to incorporate dispersive and erosive soils with a limited quantity of nondispersive, nonerodible soil to protect floodwater retention structures from flowing water, and to prevent seepage from undermining the concrete lining. The criteria used for a clayey soil if it is to be used for protecting a dike from flowing water, are that the soil should be nondispersive, should have a PI greater than or equal to 10 percent, and should contain sufficient gravel for erosion resistance. The design, therefore, had severe material property constraints and limited quantities of materials that met these design requirements. In addition, the majority of the earthen structures were to be composed of erosive soil that had to be protected.

The final design addressed all design concerns previously mentioned in the beginning of this section. Figure 3 gives a typical cross section of the canal. The typical section shows the compacted material used to cut off pervious sand lenses and layers, when encountered; as well as the overexcavation requirements, when required.

In the final design, the reservoir side of the collective dike was composed of the best material available, which was available only in limited quantities from the borrow area. The material was placed dry of optimum moisture content (compacted at 12,375 ft-lbf/ft^3) to minimize the possibility of desiccation cracks. This material, termed zone 1 material, was nondispersive, had a PI equal to or greater than 10, and contained a minimum of 10 percent fines. The requirement for a PI greater than or equal to 10 was to assure at least a minimal amount of soil plasticity for erosion resistance and healing capability upon rewetting should desiccation cracks occur. The requirement for 10 percent fines was to assure a water-resistant structure.

This material was placed on the upstream face and over the top of the compacted cutoff and core material for protection from laterally moving surface water parallel to the upstream face of the collective dike. The zone 1 material on top of the dike was used to protect the core material from the erosional effects of rainwater entering possible desiccation cracks in the core.

Zone 2 material, which was nondispersive and had a minimum of 10 percent fines, formed the cutoff and the bulk of the compacted embankment for the collective dike. The cutoff was designed to intercept pervious lenses or layers which, if permitted to carry water, might cause uplift beneath the concrete lining when the canal is dry. Both the zone 1 and zone 2 materials were compacted to specified dry unit weight and moisture content requirements.

Zone 3 material formed the remainder of the collective dike. The waste bank was also composed of zone 3 material. This material was composed of all material not utilized as zone 1 or zone 2, i.e., excavation from the canal which could not be used because it was dispersive or contained less than 10 percent fines. Zone 3 did not have specified compaction requirements, but was wetted and haul traffic was directed over it.

Transverse windrows were constructed approximately every 200 feet on top of the collective dike and waste bank to collect and slow the movement of rainwater along the dike, to pond the rainwater in an attempt to heal desiccation cracks, and to provide water for desert vegetative growth on top of the dike.

The distance between the collective dike and the centerline of the canal was controlled by the anticipated volumes of dispersive and waste

Distance between ℄ of Aqueduct and ℄ Collective Dike

STATION	TO STATION	FT.
110+00	214+00	110.0
214+00	282+00	130.0
282+00	345+00±	150.0

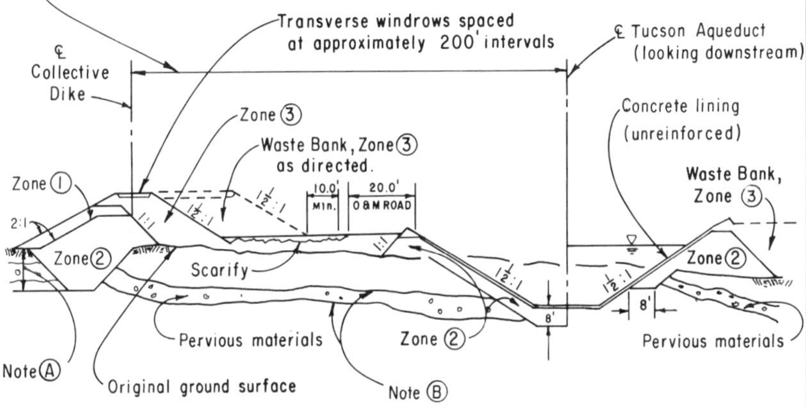

ZONE MATERIAL REQUIREMENTS

Zone ① The compacted embankment material shall be nondispersive, shall have $PI \geq 10$, shall contain a minimum of 10 percent fines and shall be obtained from the designated borrow site.

Zone ② The compacted embankment material shall be nondispersive and shall contain a minimum of 10 percent fines.

Zone ③ The embankment material and waste bank shall consist of all materials not used in Zone 1 or Zone 2.

Note Ⓐ: Depth of the cutoff trench shall equal the distance from the original ground surface to the top of Zone 1 in the collective dike, except that the bottom of the cutoff trench shall not go below the canal invert.

Note Ⓑ: If the interface between the dispersive clays and pervious sands and gravels is non-existent or is more than 5 feet below the canal invert, then overexcavation around the canal will not be required below the top of the concrete lining.

FIGURE 3 — TYPICAL SECTION — TUCSON AQUEDUCT, REACH I

material from canal excavation. These volumes, in turn, were based on dispersive testing conducted on samples at 5-foot depth intervals from the drill holes and auger holes.

Overexcavation beneath the canal addressed the concern about seepage from the canal undermining the lining by piping dispersive soil from the foundation into underlying pervious layers. Zone 2 material was recompacted into the overexcavated area. The overexcavation requirement shown on figure 3 was used to keep canal leakage from passing through the dispersive soil and into the pervious material below. It was anticipated that this problem would occur only locally, as indicated on only a few auger holes, and that most of the canal alignment would not require overexcavation. Inspectors and geologists were required to monitor the excavation daily to see if this assumption was correct; as discussed later, it was not.

Prior to placement of embankment material, the foundation was prewetted by sprinkler irrigation to a depth of 5 feet below the original ground surface or to impervious material, whichever was less. The foundation was then stripped if unsuitable materials were encountered, such as loose sands in arroyos, and scarified to a depth of at least 6 inches underneath zone 1 and zone 2 materials and 8 inches underneath zone 3 material. This was done in an attempt to minimize differential settlement and to assume a good bond between foundation and embankment material.

Prior to excavation, prewetting was accomplished by sprinkler irrigation to a depth of 8 feet below the canal invert, 8 feet below the invert of the cutoff trench, 5 feet below any overexcavation line, or 5 feet below original ground surface in areas when embankments composed of zone 3 material were constructed. The prewetting prepared the material for placement as either zone 2 or as zone 3 material and controlled dust, as required by law. The foundation material was also wetted in an attempt to densify under load any low density areas.

CONSTRUCTION PROBLEMS

The design attempted to address the geologic and design concerns. However, problems occurred in implementing the design and difficulties were encountered by the field in attempting to adhere to the specifications.

The contract to construct Reach 1A, Tucson Aqueduct, was awarded in May 1984. Excavation began in the summer of 1984. During excavation of the canal prism, it was discovered that the pervious sand and gravel lenses and layers underneath the canal invert were more extensive than expected from the exploration program. Previous geologic reports had mentioned the possibility of thin accumulations of clean sands in drainages, but these deposits were not extensive enough to warrant mapping. Now, the USBR was faced with an estimated additional cost of $250,000 to overexcavate and compact material beneath the invert for the first 4,000 feet of canal excavated. The typical section detail required that the canal prism be overexcavated 8 feet on both sides measured horizontally in areas of dispersive soil overlying pervious lenses. Further, additional work on remaining Reach 1 and on Reaches 2 and 3, which was soon to begin, would add additional cost if the same problems were encountered, as was likely to be the case. Thus, the designers and field engineers were faced with the problem of meeting design requirements but still producing an economical design. Five possible solutions were considered:

A. Excavate the canal prism without regard to overexcavation limits. Where lenses of pervious materials were found, remove any large continuous lenses but ignore the smaller lenses if they appeared to be small pockets of well graded sand; then place compacted material according to the typical section.

B. Excavate the canal prism without regard to overexcavation limits. Where lenses of pervious materials were found, remove all individual lenses with a backhoe according to the typical section; then place compacted material as specified.

C. Excavate lenses with a backhoe sufficient to provide a 2-foot-thick band of compacted material measured normal to the lining. The replacement material would be compacted to the same unit weight as the surrounding foundation material, which may be less than the specified unit weight.

D. Overexcavate the canal prism where large numbers of small lenses of pervious materials were found close together; remove all materials, both good and bad, with large equipment. The limits of the area to be excavated would depend on the typical section and the operating efficiency of the equipment. Then place compacted material according to the typical section.

E. Within the entire limits of dispersive material, require overexcavation to the limits of the typical section without regard to the number of pervious lenses found; then place compacted material according to the typical section. This solution might remove more material than required but would be easier to administer than the other solutions.

During a meeting called to resolve this problem, the design personnel emphasized the importance of a good foundation for the canal lining and would support either solutions B, C, D, or E. Solution C would be suitable, and the easing of compaction requirements would limit the amount of extra work required of the contractor.

The project office staff decided that the specifications were acceptable and did not need revisions in order to handle the pervious lens problem. The canal prism was to be excavated first; and if pervious materials were encountered, a decision was to be made at the project as to which of solutions (B, C, or D) should be followed.

These criteria were indeed workable from a contractual standpoint, but more material was removed than originally anticipated. Extra cost was added, but it is anticipated the foundation will prove adequate. Unfortunately, an evaluation of the performance of the canal lining will not be possible until water has been flowing in the canal for a while, perhaps years. Water deliveries to the Tucson area are to begin in 1991.

One problem developed after construction on Tucson Aqueduct, Reach 1B, at station 555+50, where the maximum cutoff depth was 5 feet below the original ground surface, as opposed to a depth of 20 feet for Reach 1A. After a heavy rainstorm, cracking of four concrete panels, which were subsequently removed, was discovered. Excavation between the collective dike and the canal uncovered a pervious sand lens at a depth of 8 feet which had a higher than expected moisture content. Further investigations showed that the cutoff was not extended through the sand lens. Initially, sand lenses or layers deeper than 5 feet were not expected to intercept a source of surface water; obviously, at station 555+50, it did. The sand lens terminated at the concrete lining in the location where the cracking occurred. Apparently, a small head difference (about 6 inches) through the pervious materials in the area was sufficient to crack and move the concrete panels. Repairs included contour ditches and filling to minimize the time that water would stand behind the collective dike. One design assumption on Reach 1A was thus proven correct, which was that the pervious lenses could carry enough water to cause damage whether due to hydrostatic pressure against the lining or because the lenses were capable of sufficiently high flows. Time will tell if other design assumptions were also correct.

SUMMARY AND CONCLUSION

An extensive geotechnical field exploration and laboratory testing program was conducted because of dispersive and erosive soils along the alignment of the canal. The canal design was based on the need to utilize these soils in the earthen structures as well as to protect the lining of the canal from loss of support through piping of dispersive soil from the foundation into pervious layers below. One design assumption was proven correct after a heavy rainfall on Reach 1B. Only operation of the canal will show if the remaining design assumptions are also correct.

ACKNOWLEDGMENT

Thom Fisher and Richard Straubinger, Design Engineers, Canals, Structures, and Roads Section, E&R Center, Bureau of Reclamation, Denver, Colorado; and Tom Gorman, Field Engineer, Arizona Projects Office, Bureau of Reclamation, Phoenix, Arizona; assisted in gathering details for this paper.

REFERENCES

Acciardi, Raymond G., Quantification of Pinhole Test Equipment Hydraulic Characteristics, REC-ERC-82-15, U.S. Department of the Interior, Bureau of Reclamation, Denver, Colorado, September 1982.

Craft, C. D., and R. G. Acciardi, "Failure of Pore Water Analyses for Dispersion," Journal of the Geotechnical Engineering Division, ASCE, vol. 110, No. GT4, April 1984.

Earth Manual, 2d edition, U.S. Department of the Interior, Bureau of Reclamation, U.S. Government Printing Office, Washington, D.C., reprint 1980.

Emerson, W. W., "A Classification of Soil Aggregates Based on Their Coherence in Water," Australian Journal of Soil Research, vol. 5, pp. 47-57, 1967.

Kinney, James L., Laboratory Procedures for Determining the Dispersibility of Clayey Soils, REC-ERC-79-10, U.S. Department of the Interior, Bureau of Reclamation, Denver, Colorado, September 1979.

McNearny, R. L., Internal Memorandum, Subject: "Soil Dispersion Test Results of Lime-Treated and Fly Ash-Treated Soil - Tucson Aqueduct, Reach 1 - Central Arizona Project, Arizona," Geotechnical Branch Memorandum Reference No. 85-70, Bureau of Reclamation, Denver, Colorado, October 29, 1985.

Project Data, U.S. Department of the Interior, Bureau of Reclamation, U.S. Government Printing Office, Denver, Colorado, 1981.

Sherard, James L., L. P. Dunnigan, and Rey S. Decker, "Identification and Nature of Dispersive Soils," Journal of the Geotechnical Engineering Division, vol. 102, No. GT4, Proceedings Paper No. 12062, ASCE, April 1976.

Straubinger, Richard, Report of Telephone Call, Subject: "Economic Evaluation of the Use of Fly Ash and Lime for Tucson Aqueduct," Bureau of Reclamation, Denver, Colorado, July 10, 1983.

Volk, G. M., "Method of Determination of the Degree of Dispersion of the Clay Fraction of Soils," Proceedings, Soil Science Society of America, pp. 432-445, 1937.

Wellendorf, Cathy S., Geologic Design Data Report for Tucson Aqueduct, Reach 1, Central Arizona Project, Bureau of Reclamation, Phoenix, Arizona, August 1982.

Whitaker, Wade, Second Supplemental Geology Design Data Report, Borrow Material Investigations for Tucson Aqueduct, Reach 1, Central Arizona Project, Arizona, Bureau of Reclamation, Phoenix, Arizona, October 1983.

USBR CASE HISTORY OF LOESS CUT SLOPES IN NEBRASKA

Larry D. Cast*
Theresa J. Casias**
Robert A. Baumgarten**

INTRODUCTION

Surficial deposits of loess blanket much of eastern and southern Nebraska (fig. 1). These deposits are a western extension of loessial deposits that formed along the Missouri and Mississippi River Valleys. It is in this material that the USBR (U.S. Department of the Interior, Bureau of Reclamation) has constructed several irrigation projects, whose main features include dams and reservoirs, and related distribution systems of concrete lined canals, earthen canals, and pipelines. During the 1950's and early 1960's, several projects were completed in loessial material of Nebraska. This early work required a considerable amount of field testing and research that resulted in

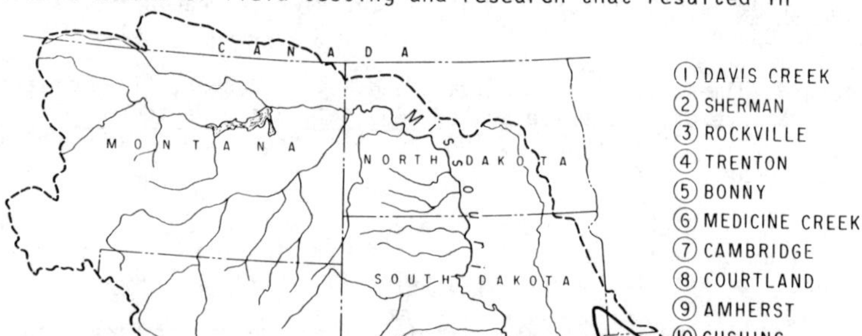

Figure 1. - Location of major loess deposits in the Missouri River Basin

*Project Geologist, Bureau of Reclamation, Grand Island, Nebraska.
**Civil Engineer, Bureau of Reclamation, Engineering and Research Center, Denver, Colorado.

preparation of the USBR Engineering Monograph No. 28, "Petrographic and Engineering Properties of Loess" (Gibbs and Holland, 1960), which became the USBR guidebook for engineering design in loess.

Evaluation of maintenance problems associated with existing canals constructed in loessial deposits of Nebraska prompted changes in USBR design criteria. These changes are reflected in recent designs for canals in the USBR North Loup Division of Nebraska.

Recent problems encountered during construction of these canals have prompted additional changes to the canal design. This paper discusses the development of current design criteria for USBR canals constructed in loessial deposits in central Nebraska.

GEOLOGIC AND ENGINEERING PROPERTIES

The North Loup Division lies in the heart of the Great Plains region. This area is typified by ten to several hundred feet of unconsolidated Pleistocene deposits overlying horizontally bedded Tertiary and Cretaceous sediments. Surficial Pleistocene deposits include silty sands of the Sand Hills of west-central Nebraska, silts and sands of the river valleys, eolian silts of southern and eastern Nebraska, and glacial tills found in portions of eastern Nebraska. Canals and laterals are located within the rolling loessial hills of central Nebraska; but approximately 20 miles (32 km) to the northwest, these loessial hills change to the Sand Hills of west-central Nebraska.

Glacial ice sheets are not recognized as having advanced into this area of central Nebraska; however, their influence was recorded by alternating periods of stream downcuttings associated with glacial advances and related lowered ocean levels, by valley fillings during glacial meltbacks, and by sediment-laden streams that moved back and forth across wide expanses of the area in Pleistocene times. Floodplains of these interglacial streams are considered to be the sediment sources of loessial deposits.

There are several recognized loesses occurring in Nebraska. Peorian Loess has widespread occurrence in the upper stratum and, consequently, is the material most frequently encountered during USBR construction activity. Peorian Loess was the subject of research and testing in the monograph by Gibbs and Holland (1960). Other loesses are older and have very limited surface exposures in Nebraska; they are generally lean clay and have undergone loading and consolidation, thus having different engineering characteristics than Peorian Loess.

Peorian Loess was deposited during the middle Wisconsin period of the Pleistocene epoch. Loess is considered to be the product of glacial-related abrasian which produced rock-powder silt that was deposited along flood plains of rivers. This was subsequently transported and redeposited by wind action. Peorian Loess is a buff colored, uniformly sorted mixture comprised predominantly of quartz grains in silt and fine sand sizes. The majority of these grains are coated with very thin films of clay. This clay is generally montmorillonite and forms intergranular supports or braces within the structure. Calcite, in loess, usually occurs as distinct silt-sized grains in a finely dispersed state rather than as a cementing material. Thin clay coatings and, to a lesser extent, calcite apparently bond particles together. Upon wetting, this bond weakens causing loss of strength.

Loess encountered during construction of the North Loup Division was a clayey to silty loess (Earth Manual, 1980) containing less than 5 percent sand (usually 1 to 3 percent) and 18 to 24 percent .005-mm or smaller sized material. According to ASTM Designation: D 2487-85 (1986 Annual Book of ASTM Standards), this loess is classified as silt (ML), silty clay (CL-ML), or occasionally lean clay (CL). The loess had a PI (plasticity index) that normally fell in the 6- to 11-percent range with an LL (liquid limit) range from 22 to 31 percent. Undisturbed dry unit weights of the loess ranged in the low 70's to low 90's lbf/ft^3 (11.0 to 14.1 kN/m^3), normally ranging between 77 and 87 lbf/ft^3 (12.1 and 13.7 kN/m^3). Maximum unit weight of the material normally ranged from 99 to 104 lbf/ft^3 (15.6 to 16.3 kN/m^3), with an optimum moisture content of 19 to 20 percent. Field moisture content of the loess was highly variable and dependent on depth of sampling, type of vegetation cover, and climatic conditions.

PERFORMANCE OF EXISTING CANALS

During the 1950's and 1960's, the USBR utilized two typical canal sections having cut slopes of 1/4:1 and 1/2:1 in loess. All canal slopes were limited to a maximum height of approximately 30 feet (9 m), at which height a bench was constructed. Bench widths ranged from 14 to 20 feet (4.3 to 6.1 m) sloping slightly toward the back of the bench. General sketches of the two typical sections are shown on figures 2 and 3.

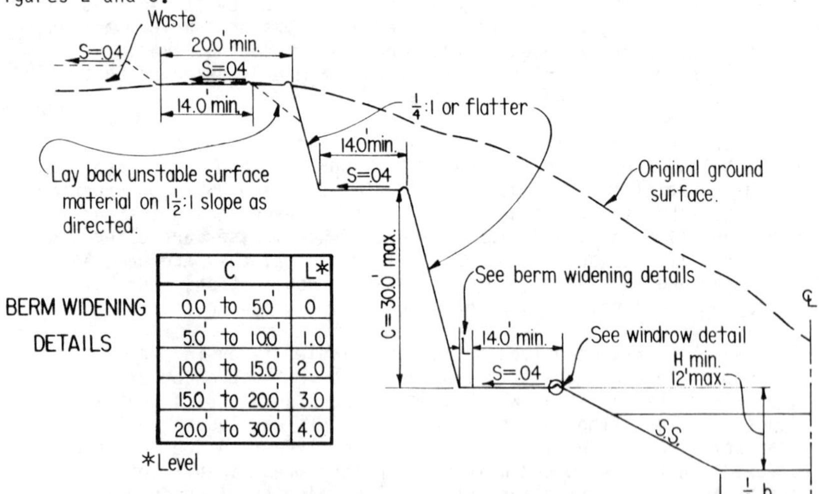

Figure 2. - Typical 1/4:1 canal cross section constructed during 1950's and 1960's

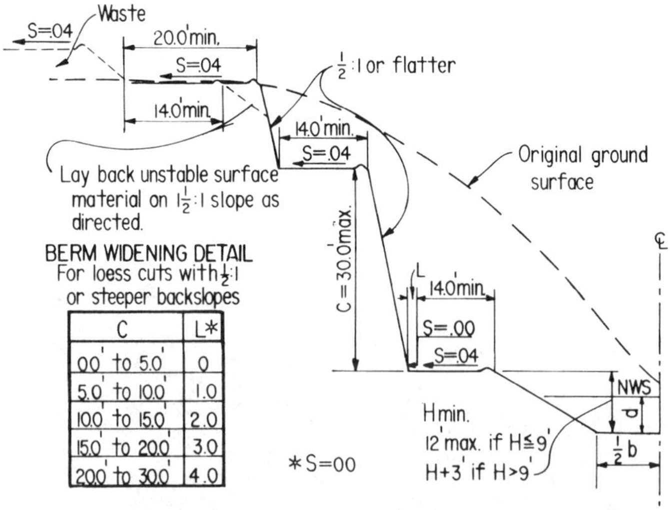

Figure 3. - Typical 1/2:1 canal cross section constructed during 1950's and 1960's

Figure 4 is a current photograph of Driftwood Canal, located in south-western Nebraska, as it exists approximately 30 years after construction. Note relative stability of the 1/4:1 slope with very little sloughing and erosion. Also, note that there is an insignificant amount of vegetation growth on the slope.

Figure 4. - Driftwood Canal (1/4:1 cut slope) approximately 30 years after construction

Figure 5 is a photograph of Sherman Feeder Canal, located in south-central Nebraska, approximately 25 years after construction. The 1/2:1 slope promotes increased vegetation growth, less erosion, and a stable condition. Also, note the torn soil fabric; this condition is frequently encountered on steep hillsides.

Figure 5. - Sherman Feeder Canal (1/2:1 cut slope) approximately 25 years after construction

Documentation detailing reasons for past designs and problems encountered during construction are limited; however, recent inspection of these cut slopes indicates that all are functioning adequately. Localized slope failures have occurred on several sections of these canals requiring continual maintenance. Localized slope failures have posed problems in maintaining O&M (operation and maintenance) roads. In sections of these canals where O&M roads are frequently traveled, fallen debris from steep cut slopes has been spread onto the tops of O&M roads. This practice has raised heights of several O&M roads by a few feet. A photograph of a raised O&M road is shown on figure 6. In many cases, fallen debris from cut slopes has created a toe at the base of the slope having roughly a 1:1 slope, as shown on figure 7.

Slopes having a thick coat of vegetation (generally slopes of 1/2:1 or flatter) do not show signs of extensive erosion (fig. 5). The presence of a good drainage system is the most crucial item in controlling erosion. Benches on cut slopes which have accumulated debris often block drainage paths, forcing water to run down the face of lower cut slopes resulting in severe erosion. Figure 8 is a photograph showing extensive erosion of a cut slope.

Figure 6. - O&M road raised with fallen debris from cut slope

Figure 7. - Natural toe formed from fallen debris from cut slope

Figure 8. - Severe erosion of cut slope due to blocked drain

NORTH LOUP DIVISION

Description

The North Loup Division is a USBR project currently under construction for the Twin Loups Irrigation and Reclamation Districts in central Nebraska. The North Loup Division is a multipurpose project (fig. 9) which derives benefits from irrigation, ground-water recharge, water quality improvement, flood control, and recreation. The main purpose of the project is irrigation. Two dams will store 136,000 acre-feet (2,484,000 hectares) of water, which will be available to irrigate 53,000 acres (21,500 hectares); 162 miles (262 km) of canals with capacities of 12 to 720 ft^3/s (0.3 to 20.4 m^3/s) will be used for water delivery. Canals having capacities greater than 50 ft^3/s (1.4 m^3/s) are constructed as open ditches, and those smaller are constructed as buried pipelines. A large percentage of these canals have been and are to be constructed in deposits of Peorian Loess. In the North Loup Division, the thickness of the loess ranges from a few feet to more than 100 feet (30 m), with an average thickness of 40 to 50 feet (12 to 15 m). Depths of canal excavations range up to 90 feet (27 m).

Design Considerations

For economic savings in reducing excavation quantities, the USBR, state and local road departments, and railroads used the unique ability of Peorian Loess to stand at near-vertical slopes. Nebraska agencies or groups involved in road building or repair have, for the most part, abandoned the use of near-vertical slopes because of aesthetics.

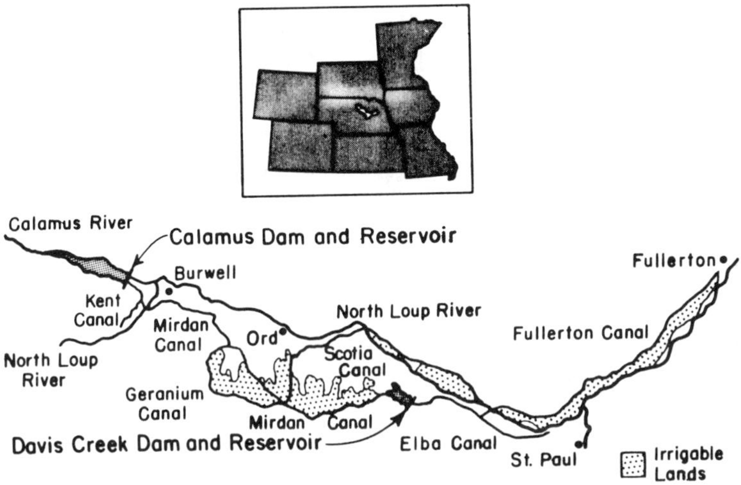

Figure 9. - USBR North Loup Division located in central Nebraska

Near-vertical slopes are slow to revegetate, and erosion-damaged faces are very conspicuous. Narrow, deep cuts in loess also serve as snow traps.

The USBR has utilized steeper slopes as an excavation cost-saving measure. Historically, the USBR used 1/4:1 and 1/2:1 slopes in Peorian Loess. When cut depths exceeded 30 feet (9.1 m), a horizontal bench was included in the slope design. Benches had widths of 14 to 20 feet (4.3 to 6.1 m), a slight slope towards the back of the bench to keep concentrated surface runoff from going down the face of the excavation, and a lateral gradient to carry water away and prevent it from collecting on the bench. In the past, this design worked well during construction but proved to be a maintenance problem.

Minor slope failure, slabbing failure, and runoff debris accumulates on these benches hampering runoff and causing water to pond. Ponded water then saturates the bench and base of the slope.

Concentrated surface runoff runs down lower slope faces causing severe erosion. Preventative maintenance was not performed on benches; therefore, conditions deteriorated and access has become impossible or extremely difficult. As a result, O&M personnel are continually battling sediment being carried onto O&M roads.

Consequently, a design decision was made to eliminate benches and to allow continuous cuts to heights of 90 feet (27 m) with 1/4:1 slopes. This decision was supported by laboratory test data, stability analyses, and the existence of comparable slopes in nature which indicated loess had sufficient strength, when dry, to stand at heights in excess of 90 feet (27 m).

SIDE SLOPES	FACTOR OF SAFETY	
	F.S.1	F.S.2
BENCHED	1.2	NOT RUN
NOT BENCHED	1.2	1.1
NOT BENCHED (Soil 2 Saturated)	1.2	1.1

SOIL PARAMETERS USED IN ANALYSIS				
Soil	Soil Class.	Wet Density (lb/ft.³)	Ø′ Degrees	C′ (lb/ft.³)
1	ML-CL	114.2	33	0
2	SM	123.2	39	0

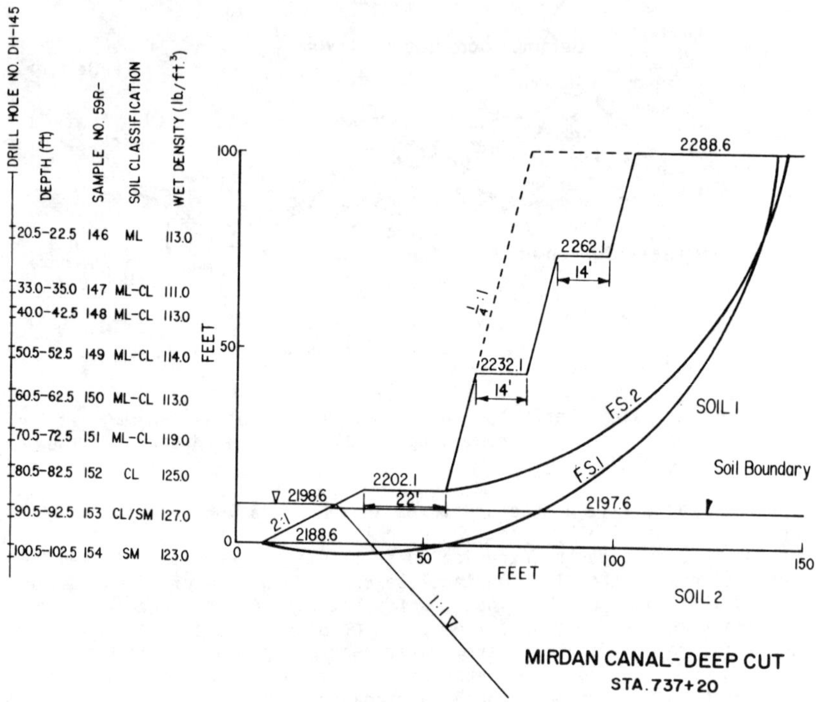

Figure 10. - Slope stability analysis for 1/4:1 cut slope

In 1981, during the design phase for Mirdan Canal, slope stability analyses were performed. Wright's modification of Spencer's method (Wright, 1974) of slope stability analysis (SSTAB1) was used to compute the factor of safety against slope failure. Soil parameters used in the analyses were obtained from CU (consolidated-undrained) triaxial shear tests performed on Pitcher tube samples obtained from the proposed canal alignment. The soil parameters used in the analyses were conservative and represent a saturated condition. The Pitcher samples were contaminated with drilling fluid, which most likely resulted in a loss of cementation upon wetting. No other sampling or strength testing of materials was performed.

Slope stability analyses showed that benching had no effect on the factor of safety against overall failure of the slope, as shown on figure 10. The possibility of localized failures in 1/4:1 and 1/2:1 cut slopes existed; however, this type of failure could not be modeled and was expected to constitute only a maintenance problem.

USBR specifications for Section 1 of Mirdan Canal included the use of unbenched 1/4:1 cut slopes through hills of Peorian Loess. Maximum height of excavated cuts above O&M roads, which parallel the canal, was to be approximately 90 feet (27 m).

Canal Excavation

Canal excavation began during the construction season of 1982. During excavation of the first major cut, a failure occurred at a depth of approximately 60 feet (18 m). This failure had a width of 100 feet (30 m), increasing to 150 feet (46 m) over a period of several days, as shown on figure 11. The exposed failure plane (head scarp) was near vertical and somewhat irregular. Along failed areas of the slope, spalling occurred at depths of 30 to 35 feet (9.1 to 10.7 m). Spalling was generally circular in nature, had a thickness of 4 to 5 inches (10 to 13 cm), and may indicate the presence of lateral forces or stresses due to stress relief. Figure 12 shows the spalling.

Figure 11. - First major cut failure on Mirdan Canal

Figure 12. - Spalling of Mirdan Canal cut slope

The head scarp exposed numerous vertical joints or cracks that were randomly oriented. Loess does contain vertical joints; therefore, it is unknown if any joints resulted from the failure. The failure is considered to have been a form of slump and toppling (Landslides - Analysis and Control, 1978). There were existing or incipient vertical joints in the loess which formed vertical discontinuities having little or no strength. When the vertical load of the wedge formed by the 1/4:1 excavated slope and the vertical joints exceeded the lateral strength of the wedge, failure or movement in a lateral or horizontal direction occurred. The block moved out laterally and then dropped down the steep face.

The remedy was to modify the 1/4:1 slope utilizing 20-foot (6.1-m) wide benches spaced at approximately 30-foot (9.1-m) vertical intervals for cut depths exceeding 35 feet (10.2 m). These guidelines were used for the remainder of the construction season. Freezing, thawing, and desiccation resulted in additional slabbing, as shown on figure 13. A decision was made to further modify slope configuration; all slopes were to be excavated at 1/2:1, and benches at 30-foot (9.1-m) vertical spacing were retained. These slopes have performed noticeably better; minor failures that have occurred were generally associated with lithologic contacts or material changes.

During the following winter and spring, blowing snow formed large drifts at the base of many of the 1/4:1 cut slopes. Subsequent melting of these drifts resulted in moisture contents close to saturation. This wet material in the 1/4:1 slopes failed in thin sheets in the lower 2 to 3 feet (0.6 to 0.9 m) of the slopes resulting in an undercut condition. A photograph of this condition is shown on figure 14. These slope failures had thicknesses of 5 to 6 inches (13 to 15 cm) and initiated upward progressive failure to the top of the

Figure 13. - Mirdan Canal slope failure on benched 1/4:1 cut slopes

slope. Debris accumulated at the toe of the slope and adversely affected longitudinal drainage along the toe. Ponded drainage further saturated the slope and finally caused transverse drainage breaking out over the bench or O&M road surface.

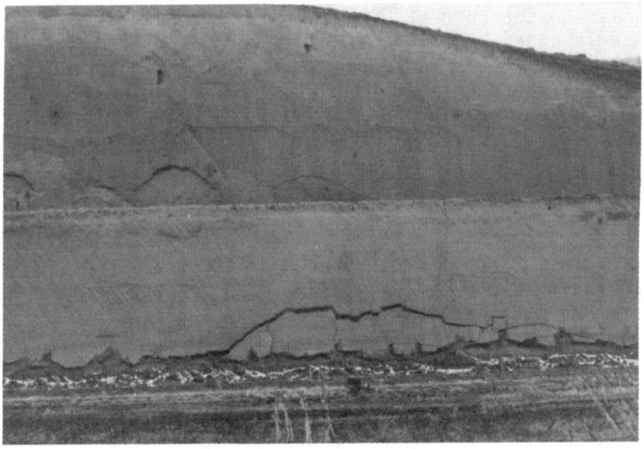

Figure 14. - Failure due to saturation along lower 2 to 3 feet of 1/4:1 cut slope

In following specifications, the USBR added a 5-foot (1.5-m) high 1:1 slope at the toe of 1/4:1 and 1/2:1 slopes. The loess is stable in a wet condition on a 1:1 slope which elimates undercutting and upward progressive failure of material. The 1:1 slope performed as expected and has been included in the design of slopes above O&M roads. These

5-foot (1.5-m) high, 1:1 slopes provide additional benefits in that they help catch debris and assist in establishment of vegetation and increase stability. Figure 15 shows the stable condition of a 1/2:1 cut slope with a 5-foot (1.5-m) high, 1:1 toe.

Figure 15. - 1/2:1 cut slope with a 5-foot-high 1:1 toe

Success of the 1:1 slope at the toe and economic preference of 1/4:1 slopes has led to the current USBR design criteria. Slopes of 1/4:1 are used when the depth or height of cut above the O&M road is less than 25 feet (7.6 m); 1/2:1 slopes without benches are used when this distance exceeds 25 feet (7.6 m). Figure 16 shows the evolution of canal slope geometry leading to current USBR design criteria.

CURRENT PERFORMANCE

There have been increased localized slope failures on canals in the North Loup Division compared to other Nebraska canals previously constructed in loessial soil. A random occurrence of limited areas of silty sands (SM) and sandy silts (ML) in the lower portion of the Peorian Loess had been identified during the exploration program, and slopes of 2:1 were specified for these intervals. However, all such occurrences were not identified prior to construction; and these unforeseen occurrences contributed to some failures of the 1/4:1 and 1/2:1 slopes.

Inspections of the recently constructed canal slopes in typical loess indicate that a 1/2:1 cut slope with a 5-foot (1.5-m) high 1:1 slope at the toe is substantially more stable. Localized failures and spalling are less prevalent in these slopes. The 1:1 slope at the toe prevents moisture from snow, rain, and construction equipment from ponding and saturating the base of the 1/2:1 cut slope. The toe also allows fallen debris from the cut slope to accumulate without interference to the O&M road and longitudinal drainage, thus reducing maintenance requirements.

16(a) -
Original design of Mirdan Canal (Section I)

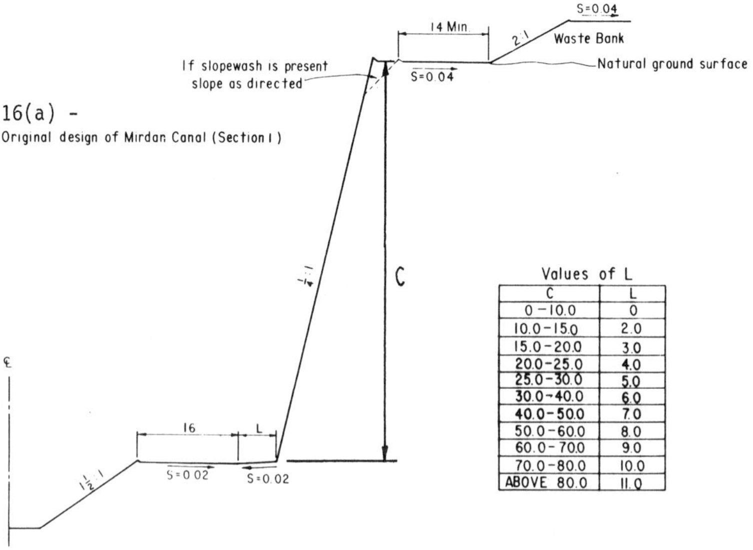

Values of L

C	L
0 – 10.0	0
10.0 – 15.0	2.0
15.0 – 20.0	3.0
20.0 – 25.0	4.0
25.0 – 30.0	5.0
30.0 – 40.0	6.0
40.0 – 50.0	7.0
50.0 – 60.0	8.0
60.0 – 70.0	9.0
70.0 – 80.0	10.0
ABOVE 80.0	11.0

16(b) -
Intermediate design during construction of Mirdan Canal (Section I)

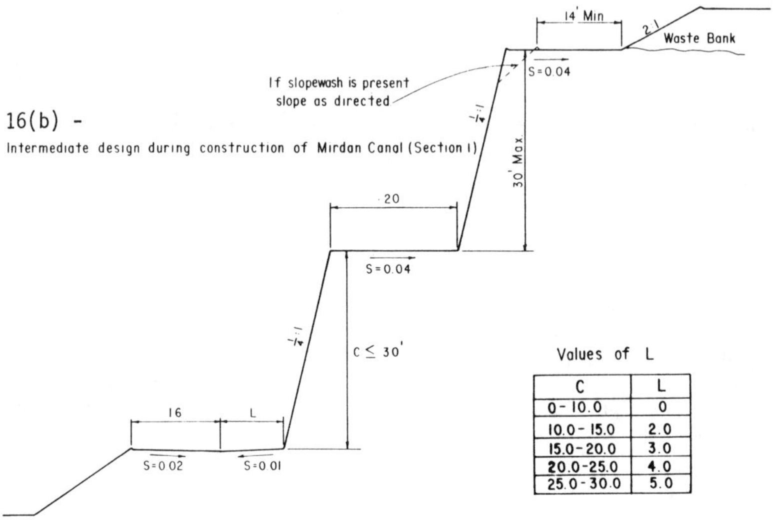

Values of L

C	L
0 - 10.0	0
10.0 - 15.0	2.0
15.0 - 20.0	3.0
20.0 - 25.0	4.0
25.0 - 30.0	5.0

Figure 16. - Evolution of canal slope geometry in the USBR North Loup Division

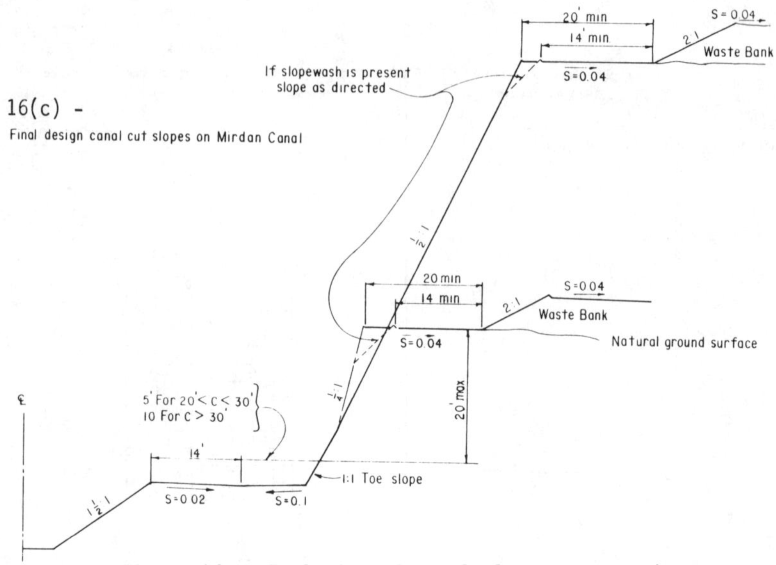

Figure 16. - Evolution of canal slope geometry in the USBR North Loup Division - Continued

Slopes of 1:1 make seeding possible and promote vegetation growth on the 1/2:1 slopes. Vegetation is important as it prevents erosion and increases stability.

With continued construction of North Loup Division canals, it has become evident that diverting drainage above the cut slopes is critical in reducing erosion and preventing slope failures by directing water away from the cut slope face. This drainage should be provided before excavation of the canal cut slopes.

PRACTICAL APPLICATIONS

Elimination of benches on 1/4:1 slopes in initial designs for North Loup Division canals has proven undesirable because of the occurrence of numerous slope failures. The benches catch debris from 1/4:1 slope failures; however, this has resulted in O&M problems because it hampers runoff and causes water to pond, saturating the bench and the base of the slope. Concentrated surface runoff also runs down the slope face causing severe erosion.

An unbenched 1/2:1 slope with a height of 55 feet (17 m) was completed during construction of Geranium Canal in October 1986. To date, no slope failures have occurred, and there is no evidence of stress in the cut slope.

Use of 1/4:1 and 1/2:1 slopes provides an excavation cost reduction over use of shallower slopes. During construction, there is no appreciable difference in excavation savings between benched 1/4:1 and unbenched 1/2:1 slopes. Introduction of the 1:1 toe and an increase in the width of the drainage berm at the base of the slope offsets excavation savings gained by elimination of benches. Primary benefits derived from elimination of the benches are reduced short-term and long-term maintenance costs. Use of 1/2:1 slopes for heights exceeding 20 feet provides an increase in vegetation growth and a decrease in slabbing and minor failures. A 1:1 slope at the toe of all 1/4:1 and 1/2:1 slopes promotes increased vegetation growth and eliminates slope failures associated with temporary saturation of the slope toe. These factors all contribute to reduced O&M costs.

CONCLUSIONS AND RECOMMENDATIONS

1. Historically, the USBR used 1/4:1 and 1/2:1 cut slopes for canals in Peorian Loess. When cut depths exceeded 30 feet (9.1 m), a horizontal bench was included in the slope design. This design worked well during construction but proved to be an O&M problem. Ponded water saturated benches and bases of slopes. Collected surface runoff ran down lower slope faces, causing severe erosion of the face. Temporary saturation at the slope bases contributed to minor failure of slope faces.

2. Recent design criteria for canals constructed in the North Loup Division allowed continuous cut heights of 90 feet (27 m) with 1/4:1 slopes. Laboratory test data, stability analyses, and the existence of comparable slopes showed loess has sufficient strength, when dry, to stand at heights in excess of 90 feet (27 m). Localized failures were possible but were only expected to create minor maintenance problems.

3. In 1982, localized failures of the 1/4:1 recently constructed slopes resulted in design changes. Benches were added to slopes with heights exceeding 35 feet (10.7 m). Benches were 20 feet (6.1 m) wide, spaced at approximately 30-foot (9.1-m) intervals. Winter freeze, thaw, and desiccation resulted in additional slabbing. Many of these failures resulted in failure surfaces having slopes of approximately 1/2:1.

4. Additional modifications were required. The slopes were excavated at 1/2:1 and benches retained at approximately 30 feet (9.1 m) vertical intervals. These slopes performed better with minor failures occurring at lithologic contacts or material changes.

5. In subsequent specifications, the USBR added a 5-foot (1.5-m) high, 1:1 slope at the toe for 1/4:1 and 1/2:1 cut slopes. The 1:1 slope at the toe has proven to be a successful addition to the design. The 1:1 slope at the toe eliminates undercutting and upward progressive failures associated with temporary saturation of the base of the slope, catches debris from above, and promotes vegetation growth.

6. Success of the 1:1 slope at the toe of the slope and economic preference for 1/4:1 slopes have led to current USBR loess design

criteria. Slopes of 1/4:1 are used when heights of cuts above the O&M roads are less than 25 feet (7.6 m); 1/2:1 slopes are used when this distance exceeds 25 feet (7.6 m).

7. During construction of North Loup Division canals, it became evident that directing water away from the cut slope face before excavation and away from the base afterwards was critical in reducing erosion and preventing slope failures. Diversion drainage is now provided before excavation of cut slopes.

8. Increased localized slope failures of canals in the North Loup Division appear to be the result of more sand lenses and localized higher sand content of the loess.

9. Monitoring ongoing excavation of 1/2:1 slopes having a height greater than 50 feet (15.2 m), with no benches, and a 5-foot (1.5-m) high 1:1 slope at the toe will continue to provide feedback on the structural feasibility of this concept. Economic benefits gained from elimination of the benches will be primarily due to short-term and long-term maintenance costs.

10. Design is an ongoing process; additional changes are continually being made to increase the efficiency and performance of USBR projects constructed in loessial soil.

REFERENCES

1986 Annual Book of ASTM Standards, sec. 4 Construction, vol. 04.08 Soil and Rock; Building Stones, American Society for Testing and Materials, 1986.

Earth Manual, 2d edition, U.S. Department of the Interior, Bureau of Reclamation, U.S. Government Printing Office, Washington, D.C., reprint 1980.

Gibbs, H. J., and W. Y. Holland, _Petrographic and Engineering Properties of Loess_, Engineering Monograph No. 28, Bureau of Reclamation, Denver, Colorado, November 1960.

Landslides - Analysis and Control, Robert L. Schuster and Raymond J. Krizek, editors, Special Report 176, National Academy of Sciences, Washington, D.C., 1978.

Wright, Stephen G., "SSTAB1 - A General Computer Program for Slope Stability Analyses," Research Report No. GE-74-1, Department of Civil Engineering, University of Texas at Austin, August 1974.

ACKNOWLEDGMENT

The authors wish to acknowledge Thomas R. Haider, Supervisory Civil Engineer, Canals and Diversion Structures Section, in the Division of Dam and Waterway Design, and Richard A. Young, Technical Specialist, Geotechnical Branch, Division of Research and Laboratory Services, for their review of this paper.

PORE STRUCTURE OF LOESS USING MERCURY INTRUSION POROSIMETRY

by

Alan J. Lutenegger[1], M.ASCE and Robert T. Saber[2], A.M.ASCE

ABSTRACT: Results of mercury intrusion porosimeter (MIP) tests are presented for a number of low clay content, high porosity loess soils form the midcontinental U.S. The pore-size distributions (PSD) obtained indicate that the pore space in these soils is generally composed of continuously connected pores with a model pore size of about 5 to 10 μm. Additionally, the pore space appears to be composed completely of toroidal pores, i.e. "ink bottle" pores which are only accessible after exceeding a threshold pressure.

INTRODUCTION

The use of mercury intrusion porosimetry to study the influence of pore structure on engineering behavior of soils has been expanding recently. This may be in part related to the fact that a number of research institutions have the necessary equipment to perform the work and the equipment which is now available is easy to operate and in many cases may be automated for direct data acquisition. However, this trend may also be a result of the fact that this technique provides a direct non-destructive method for quantifying the structure or distribution of the pore space in soil and other porous materials. Numerous studies have shown the direct connection between pore-size distribution and engineering performance of soils (e.g. Ahmed et al., 1974; Garcia-Bengochea et al., 1979; Reed et al., 1979; Delage and Lefebvre, 1984; Juang and Holtz, 1986).

Since most engineers do not use a single measure to describe the solid phase of soils but describe it in terms of the grain-size distribution, mineralogy and geochemistry, it seems likewise inappropriate to describe the pore phase of soils by a single quantity, e.g. void ratio or porosity. The arrangement of pores within a given soil sample should be expected to vary much in the same way that grain size varies, i.e., all soils do not possess a single pore size but have a range of pore sizes, referred to as pore-size distribution (PSD). Again, it is logical to assume that the PSD of most natural soil materials is a function of a number of important parameters such as geologic origin, stress history, grain-size

[1]Assoc. Prof. of Civil and Environmental Engineering, Clarkson Univ., Potsdam, N.Y.

[2]Geot. Engr., Gannett Fleming Geot. Engineers, Harrisburg, Pa.

distribution, mineralogy, etc. Thus the void ratio, which is a useful term in describing the summation of individual domain pore spaces within a sample, falls short of providing sufficient information for a quantitative measure of soil fabric. This is demonstrated by the fact that two soils may exist at the same void ratio, but have grossly different pore-size distributions. One should expect that any soil sample may have local microscopic domains which have void ratios both larger and smaller than the calculated mean void ratio for the macro sample.

As part of a project being conducted to investigate the collapse mechanism in meta-stable loess, investigations of the PSD of several loess samples were conducted. The purpose of this paper is to present the results of these investigations.

MERCURY INTRUSION POROSIMETRY

The technique of mercury intrusion porosimetry (MIP) to measure the PSD of soils makes use of the Washburn (1921) equation which states that the pressure required to force a non-wetting fluid, in this case mercury, into a cylindrical capillary pore is a function of the surface tension of the fluid, the contact angle between the fluid and the material surface, and the diameter of the pore, such that:

$$P = -\frac{4T\cos\theta}{d} \qquad (1)$$

where P = pressure
d = pore diameter
θ = contact angle
T = surface tension

The application of MIP to soils requires some concessions to be made from the ideal situation in that most soil pores are probably not cylindrical and therefore an "equivalent or apparent" cylindrical pore diameter is measured and the true shape of the pore opening is not assessed. Additionally, since most soil pores would not generally be considered to have a uniform cross sectional shape, MIP only provides a measure of the pressure required to force mercury into the accessible pore opening or pore throat. Therefore, in reality, only "equivalent pore opening" size is measured. This is often termed the "ink bottle" effect and would have the greatest influence on test results when the pore body to pore throat diameter ratio is very large.

In operation, the technique is rather simple to perform. A small sample of soil (usually < 100 mm^3) is placed in a volume calibrated sample holder (penetrometer), and a vacuum is applied to reduce the pressure to near absolute zero. After evacuation, mercury is introduced into the penetrometer to surround the sample. By measuring the volume of mercury which continues to enter the penetrometer with increasing pressure, Eq. 1 may be solved to give a record of the intruded volume as a function of the equivalent pore diameter. This record is referred to as the pore-size distribution curve and is usually expressed in terms of cumulative intruded volume versus equivalent pore diameter, Fig. 1a. An alternative may be to show differential or incremental intruded pore volume, as shown in Fig. 1b. A recent review of MIP for use in soils and other porous materials has been presented by Winslow and Lovell (1981). The reader may wish to review this work for more specific details on the technique.

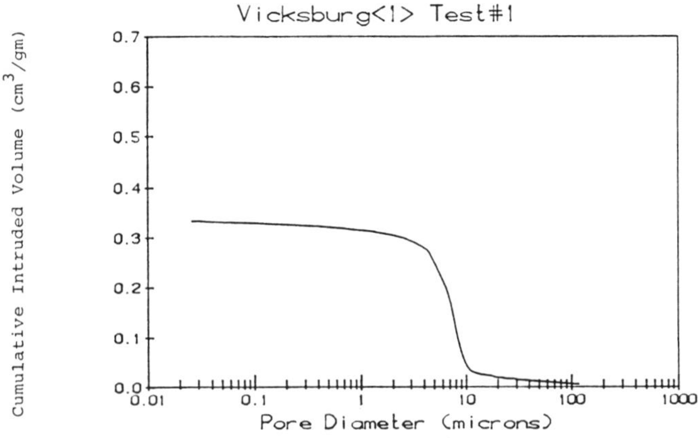

Fig. 1a. Cumulative intruded pore volume versus equivalent pore diameter

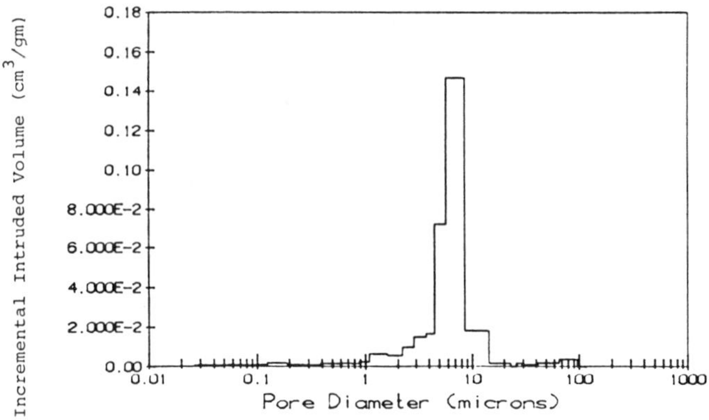

Fig. 1b. Incremental intruded pore volume versus equivalent pore diameter

TESTING PROGRAM AND RESULTS

SOILS - The pore-size distribution of a number of loess samples was investigated using MIP. Undisturbed hand-carved block samples of loess were obtained at several sites throughout the Mississippi and Missouri River drainage basins, shown in Fig. 2. All loess samples were obtained from late Pleistocene aged (Peorian) loess deposits. A summary of physical properties of samples tested is presented in Table 1.

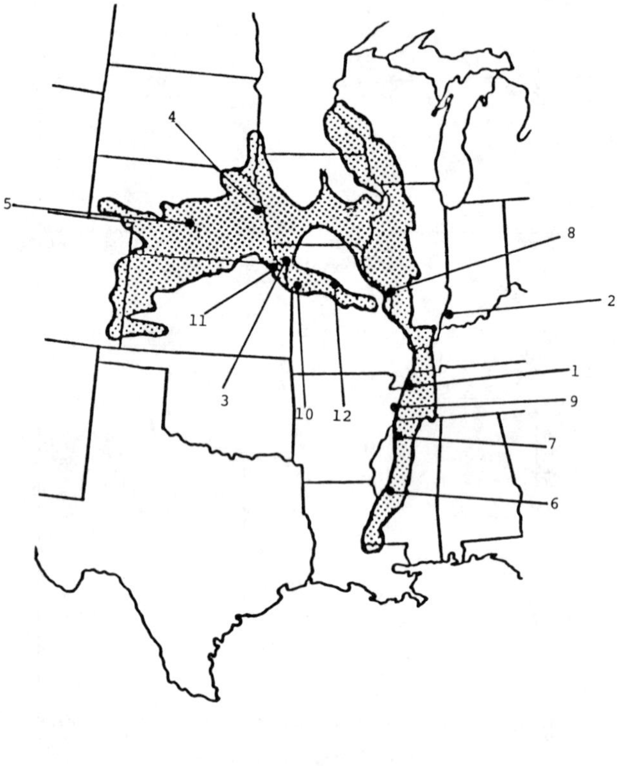

Fig. 2. Location of hand-carved loess samples. (Numbers refer to sample numbers in Table 1.) (Stippled area denotes extent of major loess deposits in midcontinent.)

Table 1. Characteristics of Loess Soils Tested

Sample	Dry Density gm/cm^2	Moisture Content %	Initial Void Ratio	Clay Content (< 2 μm) %
1 - Dyersburg, Tenn.	1.31	1.5	1.030	12.2
2 - New Harmony, Ind.	1.44	0.5	0.879	3.1
3 - St. Joseph, Mo.	1.32	2.2	1.044	15.1
4 - Omaha, Ne.	1.26	2.9	1.137	10.1
5 - Lexington, Ne.	1.39	2.7	0.944	16.2
6 - Vicksburg, Miss.	1.40	0.8	0.922	7.1
7 - Fann. Creek, Miss.	1.51	1.1	0.782	14.3
8 - E. St. Louis, Ill.	1.43	1.0	0.885	4.5
9 - Crowley Ridge, Ark.	1.48	1.4	0.828	10.9
10- Kansas City, Mo.	1.55	2.8	0.742	19.7
11- Doniphan Co., Ks.	1.42	1.9	0.892	16.6
12- Miami, Mo.	1.36	2.3	0.991	14.9

METHODS - PSD curves were obtained using the procedures outlined in ASTM D4404-84 (ASTM 1986). An Aminco 0-21.7 MPa porosimeter was used for each test. Incremental stress-controlled tests were used and a total intrusion time of about 30 min. was used for all tests. For a number of samples, a double intrusion technique (Pellerin, 1979) was used to obtain a measure of the amount of trapped mercury in individual pore ranges. This involved releasing the pressure from the mercury, allowing mercury to exit the specimen at atmospheric pressure and then reintruding the sample in a manner identical to the initial intrusion. One reason for using a reintrusion technique rather than measuring the unloading hysterisis is that the contact angle θ in Equation 1 can be considered the same in successive intrusions.

REPRODUCIBILITY OF RESULTS - The authors were initially concerned that since a relatively small specimen was being tested, it might be difficult to obtain representative results of the PSD from a single test. This might be particularly true for high porosity samples which might contain surface or internal "imperfections" or which might have a higher probability of having less uniform point-to-point porosity. In order to investigate the reproducibility of pore-size distribution, a sample of loess from near Vicksburg, Miss. was repeated a number of times. A single hand-carved sample was subdivided into smaller specimens during the course of the testing program. The results are presented in Fig. 3 for PSD tests performed on four oven-dried specimens. These results indicate that within the limitations of the technique, there appears to be excellent reproducibility of the PSD curve in this material.

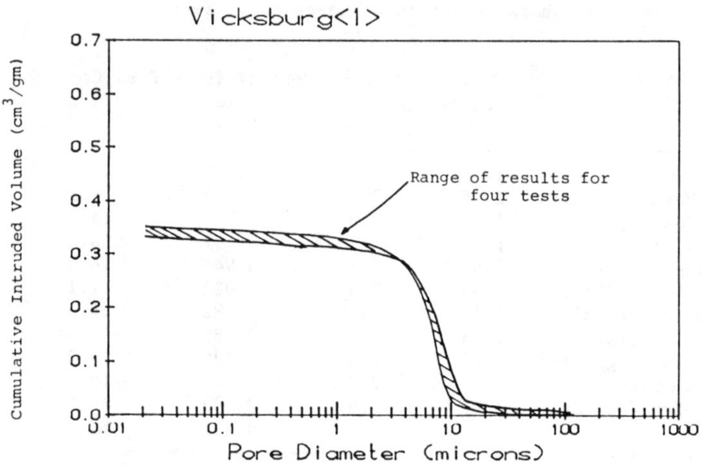

Fig. 3 Reproducibility of PSD tests for a Vicksburg loess sample

SAMPLE PREPARATION: FREEZE-DRYING VERSUS OVEN DRYING - One requirement of the testing technique is that the moisture must be removed from the pore space prior to intruding mercury. Most investigators (e.g. Ahmed et al., 1974; Delage and Lefebvre, 1984) have suggested that either critical-point drying or freeze-drying techniques be used to reduce changes in the fabric resulting from removal of the pore moisture. This would be particularly important for clayey soils which would have a high affinity for holding water; however, it appears to be a function of not only clay content and clay type, but also of initial moisture conditions (Murray and Quirk, 1980; Zimmie and Almaleh, 1976).

For the current investigation, the authors desired to reduce sample handling and preparation and overall testing time; therefore, it was decided to investigate the possibility of oven-drying loess samples prior to porosimetry work. In order to determine if oven-drying had a significant influence on the resulting pore-size measurements, duplicate specimens of several samples were tested after both freeze-drying and oven-drying. For freeze-drying, specimens were immersed in liquid nitrogen for 15 sec. and then immediately transferred to a vacuum sublimation unit. An acetone-dry ice slush was used to condense moisture during sublimation. All specimens were held in the sublimation unit for a minimum of 24 hours.

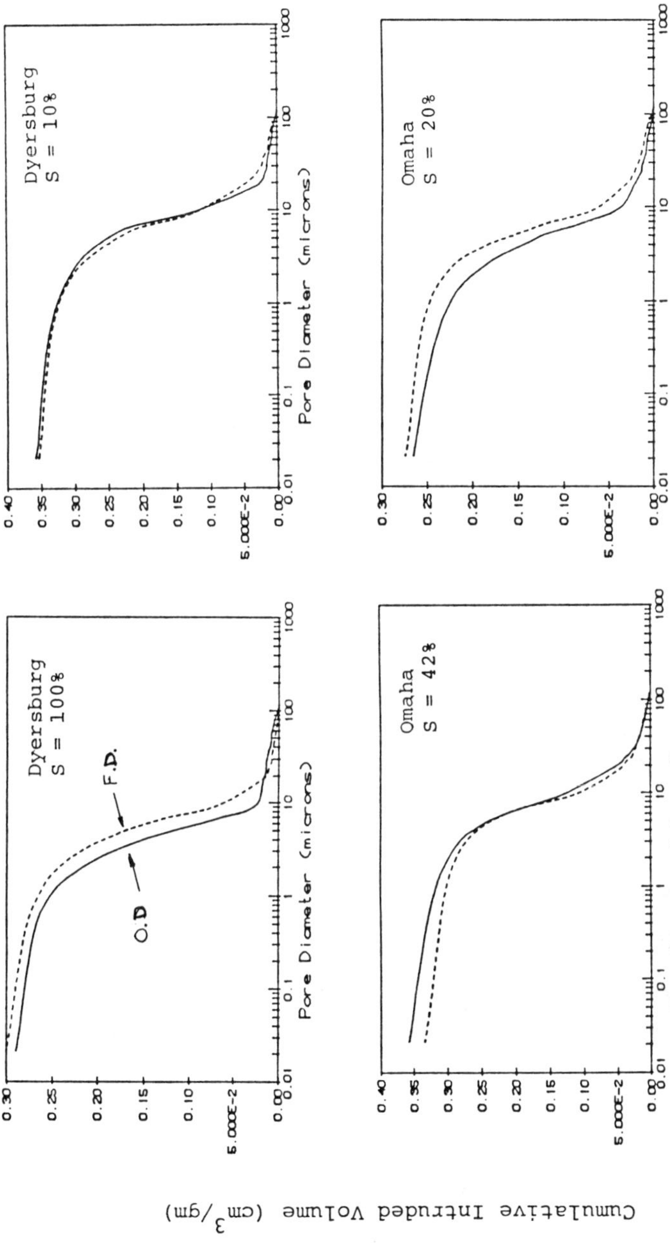

Fig. 4. Comparison of effect of oven-drying versus freeze-drying on PSD

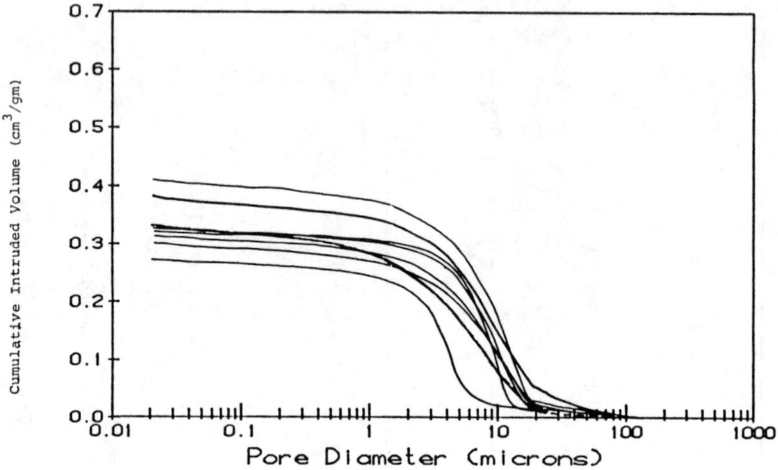

Fig. 5. PSD results for a number of loess samples from the midcontinent

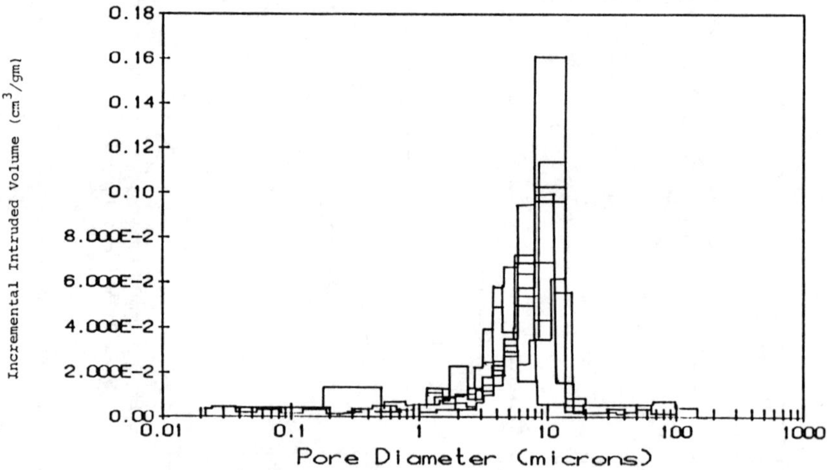

Fig. 6. Incremental PSD results for a number of loess samples from the midcontinent

Results of four typical comparison tests are presented in Fig. 4 for a range of initial moisture conditions from air dried to saturated. As noted, the PSD curve is reasonably well preserved for all samples and it would appear that oven-drying has minimal effect on the measured PSD in these soils. This is distinctly different than results obtained for clayey soils.

PORE STRUCTURE OF NATURAL SAMPLES - The pore-size distribution of a number of oven-dried loess samples obtained from around the midcontinent was determined. All samples had clay contents (<2 μm) less than 20% and were obtained from locations near the river valley source area. The mean void ratio of all samples was 0.914 ± 0.114 (the plus or minus value indicating one standard deviation). These materials would traditionally be referred to as "friable" loess. The results of these tests, shown in Fig. 5, indicate that these materials have remarkably unique PSD characterized by the majority of pores in the range of 5-15 μm. These curves compare well with a PSD curve for friable Iowa loess presented by Badger and Lohnes (1972). Additionally, within the range of pore sizes measured, the results indicate that these materials show a near-normal distribution of pore sizes, as shown by the incremental intrusion curves, Fig. 6. Furthermore, by comparing the amount of intruded volume with the calculated void volume of individual samples, the results of MIP tests indicate that there is generally less than 1% unintruded pore volume in all samples. Therefore, it would seem that the pore space in these materials is generally interconnected, and few closed pores exist.

The results of double intrusion, shown in Fig. 7 for a typical test, indicate that nearly all of the pores are "ink bottle" in configuration, i.e. nearly 100% of the mercury was trapped and did not extrude upon release of pressure. This is in contrast to finer

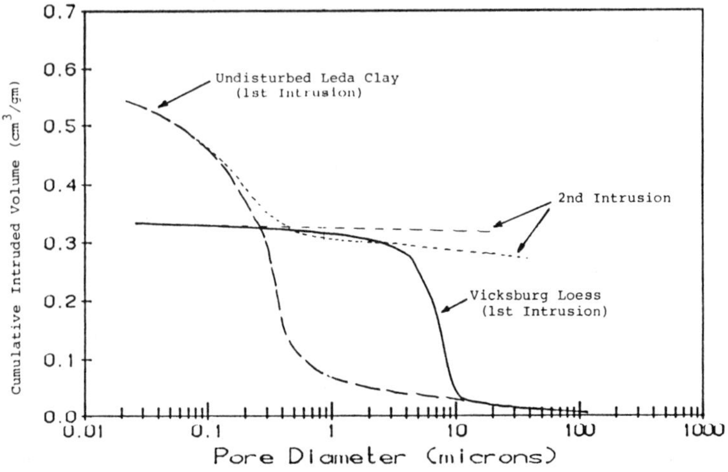

Fig. 7. Typical results of double intrusion tests

textured sediments, which show only a moderate amount of "ink bottle" pores, Fig. 7. Such high volumes of trapped mercury tend to indicate that the pore space in low clay content, high void ratio loess is composed primarily of toroidal pore space, i.e., pore space located between individual particles, and therefore the concept of these materials as a particulate system should be entertained.

EVOLUTION OF STRUCTURE DURING CONSOLIDATION - In order to investigate changes which take place in the pore structure when consolidation occurs, a series of consolidation tests was performed on an undisturbed (handcarved) identical sample of loess from Dyersburg, Tenn. Tests were conducted using a fixed-ring, dead-load, lever-arm consolidometer with a specimen diameter of 63.5 mm and a height to diameter ratio of 4. Specimens were presaturated at zero applied load and a load increment ratio of 1 was used. Since this material exhibited rapid rates of consolidation, test loads were applied when the deformation dial had essentially shown no movement. Specimens were removed following loads of 306, 611, 1222 and 4890 kPa. Results of MIP tests on oven-dried specimens are shown in Fig. 8.

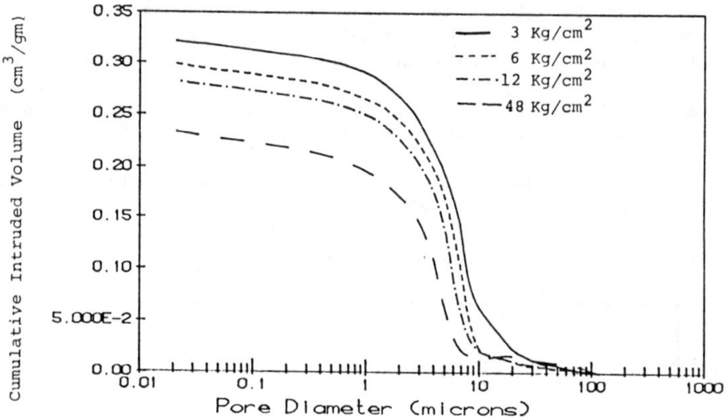

Fig. 8. PSD results from consolidated samples of Dyersburg, Tenn. loess

Admittedly, some rebound does occur when specimens are unloaded from the oedometer; however, at the present time, the authors are unaware of any procedure to hold MIP specimens at constant volume prior to testing. Therefore, the results shown in Fig. 8 may contain some unknown error in total pore volume. Nevertheless, these results show that both a reduction in total pore space (corresponding to the reduction in void ratio) and a systematic decrease in the modal (most frequently occurring) pore size is achieved by the consolidation. However, the overall shape of the PSD curve is preserved over a 16 fold increase in consolidating pressure, i.e. consolidation only causes a slight modification in the pore structure. Essentially, large voids are destroyed and new voids are created to shift the modal size from about 10 μm to about 4.5 μm. It would appear that no new voids below a size of about 2 μm are created. Therefore, the consolidation behavior of friable silts is controlled by rearrangement of the large pore space. In fact, there does not appear to be such a dramatic change, perhaps because the material is approaching a maximum particle packing arrangement. The minimum void ratio for densely packed uniform spherical particles is 0.35, which compares to a final void ratio of 0.33 for the specimen stressed to 4890 kPa.

EVOLUTION OF STRUCTURE DURING COLLAPSE - One of the goals of the work was to investigate the collapse potential of loess soils and the contribution of various size ranges of voids to the collapse mechanism. Additionally, the influence of collapse stress on the change in pore structure was of interest. A block sample of loess was allowed to air dry, and specimens were then trimmed to fit the oedometer ring. Collapse tests were conducted by loading specimens in increments of 51 kPa and then inundating with distilled water at the desired collapse stress. Tests were conducted at collapse stresses of 102, 204, 306, 407, 509, and 611 kPa. The results of collapse tests are summarized in Table 2.

Table 2. Results of Collapse Tests on Dyersburg, Tenn. Loess

Collapse Stress kPa	e_o	w %	i %
102	1.024	1.16	1.61
204	1.027	1.17	2.54
306	1.036	1.26	4.48
407	1.041	1.87	8.84
509	1.035	1.65	9.34
611	1.016	1.52	8.86

$i = \frac{\Delta e}{e_i} \times 100\%$, where e_i = void ratio at beginning of collapse = volumetric strain of collapse.

The PSD results of collapse tests showed trends similar to those obtained from consolidation tests, i.e., a reduction in modal pore diameter with increasing stress, Fig. 9, even though the stress range is substantially different. By subdividing the pore space into arbitrary ranges, the change in each range versus collapse stress may be quantified, as shown in Fig. 10. As noted, there is a significant systematic decrease in the pores 9-3 μm. All other ranges generally show little systematic change. Therefore, during collapse, the reduction in void ratio is attributed to destruction of pores of size 9-3 μm apparently and no detectable new pores are created.

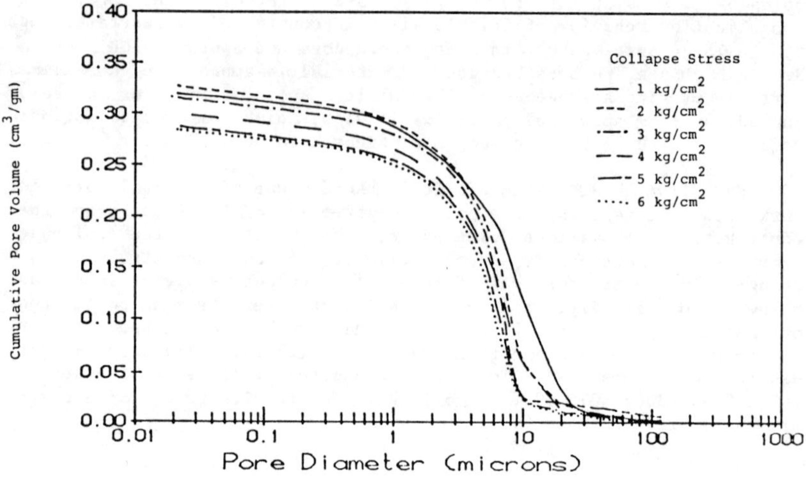

Fig. 9. PSD results from collapsed samples of Dyersburg, Tenn. loess

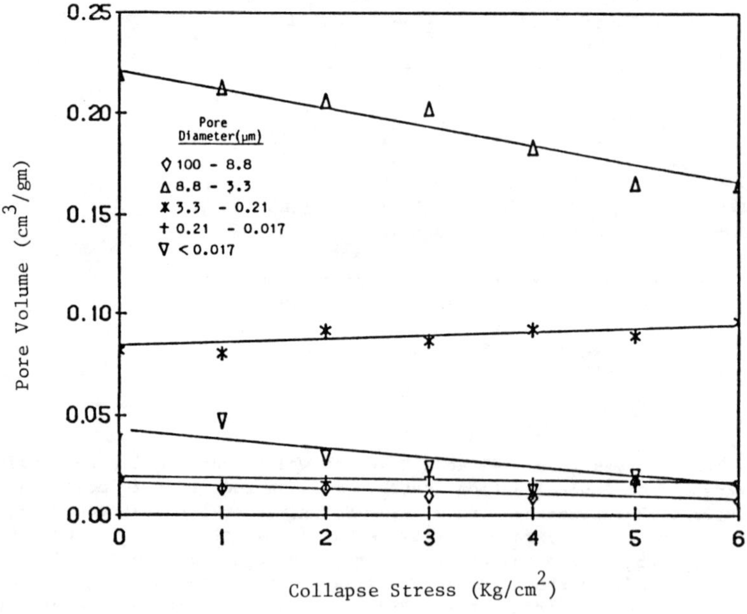

Fig. 10. Changes in pore space as a function of collapse stress

REVIEW OF RESULTS AND CONCLUSIONS

Mercury intrusion porosimetry provides a unique look at the pore structure and fabric of soils by quantification of the apparent pore-size distribution. The void ratio alone does not describe the micro-structure, but gives a combined average measure of the void space of porous materials. Individual point-to-point void ratios within natural soils reflect domain void sizes between particular groups of particles. For the loess soils investigated, the results of porosimetry tests indicate that the majority of void space in these materials is composed of toroidal pores, i.e. the pore throat or opening to pore body ratio must be relatively small since most of the mercury is trapped as indicated by double intrusion. However, it would also appear that the pore space is continuously connected since nearly all of the calculated available pore space was accessible to intrusion.

Results of porosimeter tests on natural friable loess samples from around the midcontinent show remarkable uniformity in the shape of the pore-size distribution curve. In general, the pore-size distribution approaches near Gaussian distribution, with apparent modal pore sizes in the range of 5 to 10 μm. Few small pores (< 2 μm) and few large pores (> 50 μm) exist in these soils. Changes in the pore-size distribution, induced by either consolidation or artificial collapse in the oedometer, are reflected in a loss of large pores and a reduction of the modal pore size, while the overall characteristic shape of the pore-size distribution curve is preserved.

It should be pointed out that the sharp change in slope of the PSD curves presented may be misleading and this break (which occurs at 5 to 10 μm) may actually represent a threshold pressure which must be applied to force mercury through smaller pore openings which connect to large pore bodies. This reasoning would be consistent with results of double intrusion and is currently being investigated further by multiple intrusion at various pressures.

Because of the low clay content in the samples tested, porosimeter tests on over-dried and freeze-dried samples showed essentially the same results. This may be in part due to the fact that only part of the clay fraction is actually composed of clay minerals with the remainder being clay-sized non-clay minerals.

ACKNOWLEDGEMENTS

The work reported herein was conducted under NSF Contract No. ECE-85-04460. The authors wish to thank NSF for their support in this work.

APPENDIX - REFERENCES

Ahmed, S., Lovell, C.W. Jr., and Diamond, S. (1974). "Pore Size and Strength of Compacted Clay." Journal of the Geotechnical Engineering Division, ASCE, Vol. 100, pp. 407-425.

American Society for Testing and Materials (1986). "Standard Test Method for Determination of Pore Volume and Pore Volume Distribution of Soil and Rock by Mercury Intrusion Porosimeter." Designation D4404-84, Annual Book of ASTM Standards, Vol. 04.08, pp. 862-868.

Badger, W.W. and Lohnes, R.A. (1972). "Pore Structure of Friable Loess." Highway Research Record No. 429, pp. 14-23.

Delage, P. and Lefebvre, G. (1984). "Study of the structure of a sensitive Champlain clay and of its evolution during consolidation." Can. Geotech. Jour., Vol. 21, pp. 21-35.

Garcia-Bengochea, I., Lovell, C.W., and Altschaeffl, A.G. (1979). "Pore Size Distribution and Permeability of Silty Clays." Journal of the Geotechnical Engineering Division, Asce, Vol. 105, pp. 839-855.

Juang, C.H. and Holtz, R.D. (1986). "Fabric Pore Size Distribution, and Permeability of Sandy Soils." Journal of Geotechnical Engineering, ASCE, Vol. 112, pp. 855-868.

Murray, R.S. and Quirk, J.P. (1980). "Freeze-Dried and Critical-Point-Dried Clay - A Comparison." Soil Sci. Soc. Amer. Jour., Vol. 44, pp. 232-234.

Pellerin, F.M. (1979). "La porosimetrie au mecure appliquee a l'etude geotechnique des sols et dcs roches". Bulletin de Liaison des Laboratoires des Ponts et Chaussees, Vol. 106, pp. 105-116.

Reed, M.A., Lovell, C.W., Altschaeffl, A.G. and Wood, L.E. (1979). "Frost-heaving rate predicted from pore-size distribution." Can. Geotech. Jour., Vol. 16, pp. 463-472.

Washburn, E.W. (1921). "Note on a Method of Determining the Distribution of Pore Sizes in Porous Material." Nat. Acad. Sci. Proc., Vol. 7, pp. 115-116.

Winslow, D.N. and Lovell, C.W. (1981). "Measurements of Pore Size Distributions in Cements, Aggregates and Soils." Powder Technology, Vol. 29, pp. 151-165.

Zimmie, T.F. and Almaleh, L.S. (1976). "Shrinkage of Soil Specimens During Preparation for Porosimetry Tests." ASTMTP 599, pp. 202-215. Soil Specimen Preparation for Laboratory Testing, ASTM STP 599, pp. 202-215.

DETERMINING IN-PLACE UNIT WEIGHT OF LOESSIAL SOIL

Theresa J. Casias*

INTRODUCTION

Early in 1982, USBR (U.S. Department of the Interior, Bureau of Reclamation) personnel initiated a brief investigation of push-tube sampling techniques in loessial soil. Five-inch (13-cm) diameter samples were required to provide three specimens for triaxial shear testing and 4-1/4-inch (10.6-cm) diameter one-dimensional consolidation specimens. In addition to more representative testing conditions, the larger samples were required to minimize sample disturbance, which is critical in preserving the loose in situ structure of loessial soil. In the past, five-inch (13-cm) diameter thin-wall push-tube sampling in the dry was used to obtain samples for dry unit weight determinations and laboratory testing. Radiographs (X-rays) of the samples in the tubes showed patterns of concave fracturing, perpendicular to the direction of push, in several of the loessial soil samples. Many of the concave fractures penetrated the entire sample cross section. This indicated that substantial disturbance was occurring during the push-tube sampling process. Figure 1 shows photographs of sample disturbance caused by the push tube sampling process. Further analysis of 5-inch (13-cm) push-tube sampling techniques showed that sample recovery (ratio of length of recovered sample to length of push, expressed as a percent) was low (90 percent or less) for many samples; however, it did not appear that any portion of the soil was falling out of the tube and into the drill hole. This indicated that compaction of the soil was occurring during sampling. In addition, there appeared to be some sloughing of the drill hole side walls. Because of these factors, the amount of compaction was unknown; and in-place dry unit weights obtained from 5-inch (13-cm) push-tube samples were not considered representative of in situ soil conditions. "Undisturbed" specimens for laboratory testing could not be obtained from the push-tube samples. Accurate in-place unit weights were required for slope stability analyses, and structural settlement and foundation design analyses for canal structures. Accurate in-place unit weights were also required for determining the collapse potential of the material, as collapse potential for loess is a function of dry unit weight and liquid limit (Gibbs and Holland, 1960).

Because of the problems encountered in push-tube sampling, it became necessary to investigate other sampling techniques to find a more reliable method of sampling loessial material. In May 1984, an investigation program was developed and initiated by USBR personnel to

*Civil Engineer, Bureau of Reclamation, Engineering and Research Center, Denver, Colorado.

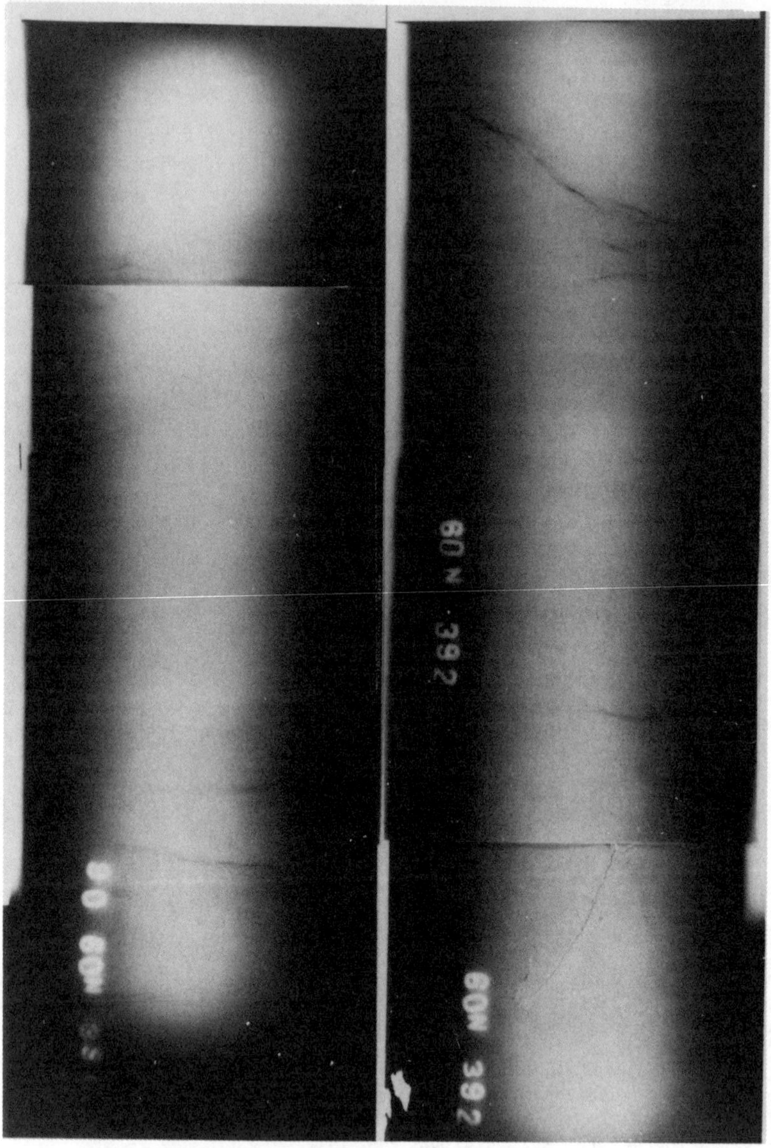

Figure 1. - Radiographic photographs showing sample disturbance during push-tube sampling

evaluate sampling techniques in loessial soil in central Nebraska (USBR North Loup Division). A detailed description of the North Loup Division and geology in the area is included in "USBR Case History of Loess Cut Slopes in Nebraska," by Larry D. Cast, Theresa J. Casias, and Robert A. Baumgarten (also included as part of this publication).

Two locations were selected for sampling along the alignment for Mirdan Canal, and two locations at Davis Creek damsite (fig. 2). The investigation program consisted of continuous sampling with a 6-1/4-inch (16-cm) i.d. hollow-stem auger system and continuous sampling with 5-inch (13-cm) i.d. push tubes in adjacent offset drill holes. In-place moisture content and dry unit weight determinations were made on samples obtained by both drilling methods. Samples were recovered from approximately the same depth intervals in adjacent drill holes. Sample recovery was calculated for all samples. A gamma-gamma downhole density logging tool was also used in each drill hole. Following completion of the geophysical logging, test pits were excavated at all four locations. Moisture content and dry unit weight determinations were made at frequent intervals in all test pits using both the Field Density Test - Sand Cone (E-24) (Earth Manual, 1980) and a surface nuclear moisture-density meter. Block samples were obtained at frequent depth intervals from each test pit for laboratory dry unit weight determinations by the mass in air - mass in water method.

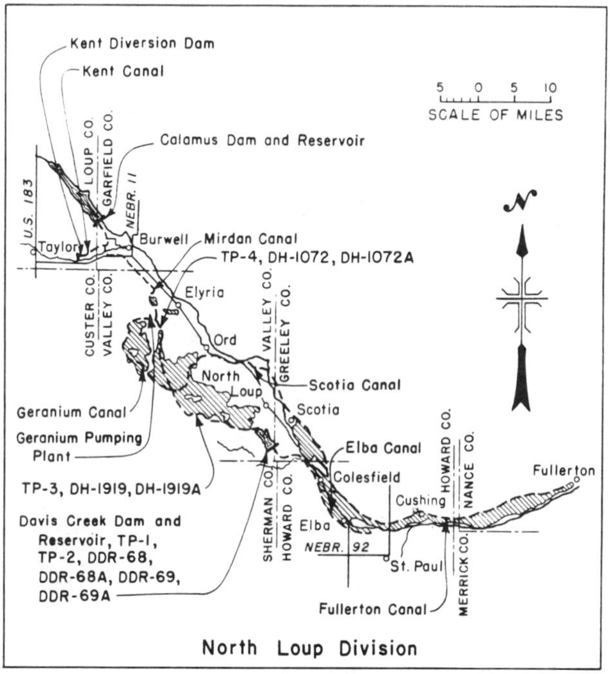

Figure 2. - USBR North Loup Division location map

Figure 3 shows a plan and profile view of a typical test pit in relation to the drill holes. Results from the sand-cone test were used as the unit weight standard for evaluating results of the other methods investigated in this program. This paper discusses results of the investigation program.

DISCUSSION

Undisturbed Sampling

Hollow-Stem Auger System

Sampling was performed using a CME-55 drill rig and a 10-1/2-inch (27-cm) o.d. flight auger with a 6-1/4-inch (15.6-cm) i.d. hollow stem, one of the few such samplers available. A photograph of the drilling equipment is shown on figure 4. A sketch of the hollow-stem auger system is shown on figure 5. During drilling operations, samples may be recovered using a core barrel sampler

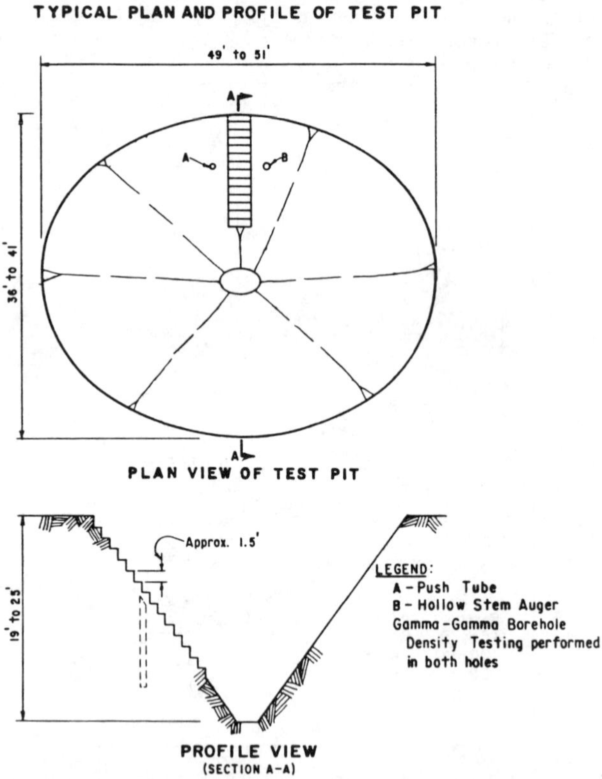

Figure 3. - Typical test pit in relation to drill holes

with or without liners; however, when samples are required for unit weight determinations, liners must be used. Sections of 5-inch (13-cm) i.d. PVC (polyvinyl chloride) pipe, cut to appropriate lengths to fit the inside of the hollow stem auger, were used as liners. A photograph of a liner made from PVC pipe is shown on figure 6.

Figure 4. - A 10-1/2-inch (270-cm) o.d. flight auger with a 6-1/2-inch (15.6-cm) i.d. hollow stem

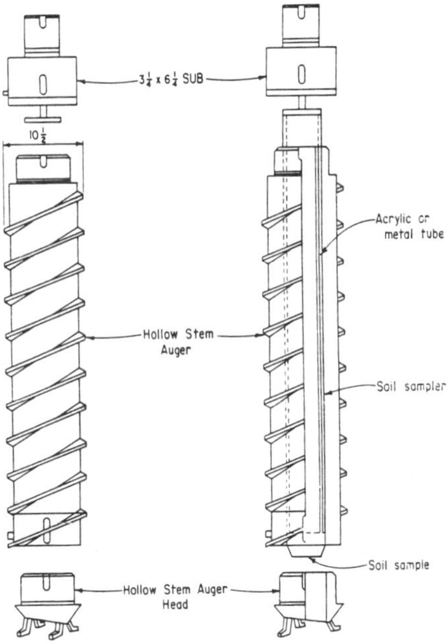

Figure 5. - Hollow stem auger (not to scale)

Figure 6. - Liner made from a cut section of 5-inch (13-cm) i.d. PVC pipe

Table 1. - Summary of in-place moisture and unit weight values - Davis Creek Dam, Pick-Sloan Missouri Basin Program

Identification			Drill hole DDR-69 Hollow stem auger				Drill hole DDR-69A Push tube			
Geologic formation (ft)	Average depth (ft)	Depth (ft)	Wet unit weight (lbf/ft^3)	Moisture content (%)	Dry unit weight (lbf/ft^3)	Sample recovery (%)	Wet unit weight (lbf/ft^3)	Moisture content (%)	Dry unit weight (lbf/ft^3)	Sample recovery (%)
0.0										
Topsoil	1.2	0.4-1.9	111.2	25.0	89.0	100				
0.0-3.7	2.6	1.9-3.4	111.0	23.3	90.1	100	118.0	24.1	95.1	73
	4.2	3.4-4.9	106.7	21.6	87.7	100	109.7	20.8	90.8	100
	5.7	4.9-6.5	96.7	23.2	78.5	100				
	5.9	4.9-6.9					100.8	22.1	82.6	100
	7.2	6.5-8.0	97.6	22.6	79.6	100				
Valley	7.7	6.9-8.5					99.9	27.9	78.1	98
Fill	9.0	8.5-9.5	99.9	24.4	80.3	100	99.4	25.5	79.2	90
	10.2	9.5-11.0					101.1	23.4	81.9	100
	10.4	9.7-11.0	96.5	20.1	80.3	100				
	11.6	11.2-12.3	102.3	20.3	85.0	100	99.4	17.6	84.5	100
	12.9	12.3-13.5	106.8	17.9	90.5	100	103.9	17.1	88.7	100
14.3	14.2	13.5-14.8					109.4	17.9	92.8	100
	14.3	13.5-15.1	109.5	17.5	93.2	100				
	15.4	14.8-15.9					107.4	19.0	90.3	96
	15.8	15.1-16.6	112.1	18.3	94.8	100				
	16.4	15.9-16.9					105.0	17.9	89.1	68
	17.6	16.6-18.5	105.2	16.5	90.3	100				
	18.6	17.6-19.5					108.8	18.1	92.1	100
	19.2	18.5-20.0	103.7	16.2	89.3	100				
Peorian	20.2	19.5-21.0					102.3	17.0	87.4	97
Loess	20.8	20.0-21.5	102.8			100				
	21.8	21.0-22.5					101.5	17.2	86.6	80
	22.5	21.5-23.5	105.5	18.4	89.1	100				
	24.5	23.7-25.3	105.3	17.9	89.3	100				
	26.0	25.3-26.6	104.9	18.1	88.8	100				
	27.4	26.6-28.1	105.5	17.9	89.5	100				
	28.9	28.1-29.7	106.1	18.0	89.9	100				
	30.5	29.7-31.1	105.9	16.6	90.8	100				
	34.2	33.4-35.0	111.8	8.8	102.8	91				
	36.0	35.0-37.0	115.7	10.6	104.6	100				

Table 2. - Summary of in-place moisture and unit weight values - Davis Creek Dam, Pick-Sloan Missouri Basin Program

Identification			Drill hole DDR-68 Hollow stem auger				Drill hole DDR-68A Push tube			
Geologic formation (ft)	Average depth (ft)	Depth (ft)	Wet unit weight (lbf/ft^3)	Moisture content (%)	Dry unit weight (lbf/ft^3)	Sample recovery (%)	Wet unit weight (lbf/ft^3)	Moisture content (%)	Dry unit weight (lbf/ft^3)	Sample recovery (%)
Topsoil										
0.0-0.4										
	2.6	1.9-3.4	97.3	20.1	81.0	100	100.3	18.8	84.4	73
	3.9	3.4-4.4	97.7	19.6	81.6	100	99.5	20.4	82.6	100
Peorian	5.2	4.4-5.9	101.4	19.3	85.0	100	109.6	19.3	91.8	83
Loess	6.6	5.9-7.4	100.6	18.0	85.2	100	102.7	17.8	87.2	100
	7.8	7.4-8.1				*				*
	8.6	8.1-9.2	100.1	16.6	85.8	100	98.2	16.5	84.3	95
	10.0	9.2-10.7	101.2	17.0	86.4	100	98.0	17.3	83.5	97
	11.4	10.7-12.1	99.2	16.9	84.8	100	98.1	16.1	84.4	100
	12.6	12.1-13.2	98.8	17.3	84.2	100	100.0	16.5	85.8	55
	13.8	13.2-14.3	99.9	17.0	85.4	100	103.6	16.0	89.3	100
	15.0	14.3-15.8	100.6	17.0	86.0	100	106.5	16.9	91.1	77
	16.6	15.8-17.3	103.8	18.2	87.8	100	111.7	16.7	95.6	77
	17.8	17.3-18.3	94.9	17.1	81.8	100	111.9	17.2	95.5	85
	18.8	18.3-19.4	99.0	19.3	83.0	100	108.1	16.7	92.7	90
	20.1	19.4-20.8	100.2	17.2	85.5	100	110.0	16.4	94.4	69
	21.6	20.8-22.3	98.4	16.1	84.7	100	114.1	16.5	97.9	83
	22.8	22.3-23.3	93.8	17.9	79.6	100	110.4	17.5	94.0	90

* Sample too small to provide representative unit weight

Table 3. - Summary of in-place moisture and unit weight values - Mirdan Canal, Pick-Sloan Missouri Basin Program

Identification			Drill hole DH-1919 Hollow stem auger				Drill hole DH-1919A Push tube			
Geologic formation (ft)	Average depth (ft)	Depth (ft)	Wet unit weight (lbf/ft^3)	Moisture content (%)	Dry unit weight (lbf/ft^3)	Sample recovery (%)	Wet unit weight (lbf/ft^3)	Moisture content (%)	Dry unit weight (lbf/ft^3)	Sample recovery (%)
Topsoil 0.0-0.8										
	3.1	2.4-3.8	96.6	20.4	80.3	100	112.9	19.5	94.5	72
	4.4	3.8-5.0	95.7	20.8	79.2	78	112.2	19.2	94.1	74
Reworked	5.7	5.0-6.4	96.5	18.6	81.4	100	112.1	18.1	94.9	72
Peorian	7.2	6.4-7.9	96.2	17.2	82.1	100	113.2	18.6	95.4	136*
Loess	8.4	7.9-8.9	99.1	17.6	84.3	100	111.9	20.0	93.2	105*
	9.4	8.9-10.0	99.7	17.8	84.6	100	111.1	19.7	92.8	95
	10.8	10.0-11.5	103.1	18.7	86.8	100				0**
	12.2	11.5-13.0	105.0	20.3	87.3	100	118.9	21.7	97.7	90
	13.4	13.0-13.8	108.3	19.9	90.4	100	118.6	21.7	97.4	106*
14.0	14.6	13.8-15.4	107.5	20.4	89.2	97	114.6	20.8	94.9	81
	15.9	15.4-16.4	105.3	19.5	88.1	100	123.4	21.8	101.3	100
	17.2	16.4-17.9	109.2	20.7	90.5	100	118.6	22.2	97.0	93
Peorian	18.4	17.9-18.9	112.2	20.6	93.0	100	116.1	21.2	95.8	105*
Loess	19.4	18.9-20.0	108.4	20.7	89.9	100	117.3	21.7	96.4	100
	20.8	20.0-21.5	108.8	21.5	89.5	100	121.4	22.0	99.5	87
	22.2	21.5-23.0	110.4	22.5	90.1	100	119.3	22.6	97.3	100
	23.5	23.0-23.9	110.4	21.0	91.2	100				

* Picked up material that caved in from side of hole
** Lost sample down hole

Table 4. - Summary of in-place moisture and unit weight values - Mirdan Canal, Pick-Sloan Missouri Basin Program

Identification			Drill hole DH-1072 Hollow stem auger				Drill hole DH-1072A Push tube			
Geologic formation (ft)	Average depth (ft)	Depth (ft)	Wet unit weight (lbf/ft^3)	Moisture content (%)	Dry unit weight (lbf/ft^3)	Sample recovery (%)	Wet unit weight (lbf/ft^3)	Moisture content (%)	Dry unit weight (lbf/ft^3)	Sample recovery (%)
Topsoil 0.0-2.9										
	2.2	1.5-3.0	94.1	22.4	76.9	100	93.3	21.6	76.7	80
	3.8	3.0-4.6	92.9	21.3	76.6	100	104.1	20.6	86.3	75
	5.4	4.6-6.1	91.7	19.3	76.9	99	111.4	20.0	92.9	77
	6.6	6.1-7.2				0*	103.5	18.9	87.1	91
	7.8	7.2-8.5	96.3	16.5	82.7	100	100.3	20.3	83.4	77
	9.0	8.5-9.5	93.7	15.0	81.5	100	115.5	18.6	97.4	100
	10.2	9.5-11.0	92.6	14.8	80.7	100	105.3	17.1	89.9	70
	11.8	11.0-12.5	93.9	14.0	82.4	100	108.4	16.1	93.4	67
	13.0	12.5-13.4					114.5	16.1	98.6	100
Peorian	14.0	13.4-14.5	94.8	14.9	82.5	100	102.7	15.6	88.9	91
Loess	15.2	14.5-16.0	95.5	14.1	83.7	100	107.9	14.7	94.1	73
	16.8	16.0-17.5	94.9	14.4	83.0	100	105.9	14.5	92.5	87
	18.0	17.5-18.5				0*				0*
	19.0	18.5-19.6	93.3	14.8	81.3	100	112.9	14.6	98.5	91
	20.2	19.6-20.9					113.2	14.7	98.7	85
	20.4	19.6-21.1	98.5	14.5	86.1	100				
	20.4	19.9-20.9				100				
	21.8	21.1-22.6	98.3	14.6	85.8	100				
	23.1	22.6-23.6				116**				
	24.4	23.6-25.1	97.0	15.2	84.2	100				

* Lost sample down hole
** Sample disturbed

Continuous hollow-stem auger sampling was performed in four holes (DDR-69 and -68, and DH-1072 and -1919). Moisture content and dry unit weight were determined for all samples, and sample recovery was computed. Sample recovery was consistently high in all samples from all four holes (see tables 1 through 4). A photograph of a typical soil sample obtained using the hollow-stem auger sampling method is shown on figure 7. Results of moisture content and dry unit weight determinations on samples obtained from all four auger holes are summarized in tables 1 through 4. Plots showing dry unit weight versus depth are shown on figure 8.

Figure 7. - Typical soil sample obtained using the hollow stem auger sampling method

Initially, sampling with the hollow-stem auger system proceeded slowly; but as the drill crew became familiar with the equipment, the sampling process became quite efficient.

One problem encountered with this sampling method was that the i.d. of the PVC pipe (liner) was slightly larger than the i.d. of the sampler bit. During the investigation program, the bit diameter was modified several times; however, the i.d. of the PVC pipe varied enough so it was virtually impossible to consistently match bit diameter to pipe diameter. An unsuccessful attempt was also made to use 5-inch (13-cm) i.d. clear acrylic tubing so a visual examination of core could be made before extruding the sample. The o.d. of the clear acrylic tubing was smaller than the i.d. of the core barrel sampler and caused the tubing to move to one side of the core barrel. Sample disturbance was then observed on one side of the sample in the clear acrylic liner. Because the PVC pipe fit more tightly in the barrel than the clear acrylic liner, this type of disturbance did not appear to be a problem with the PVC liners. Sampling then continued with PVC pipe as liners. Radiographic examinations of samples obtained using PVC pipe as a liner

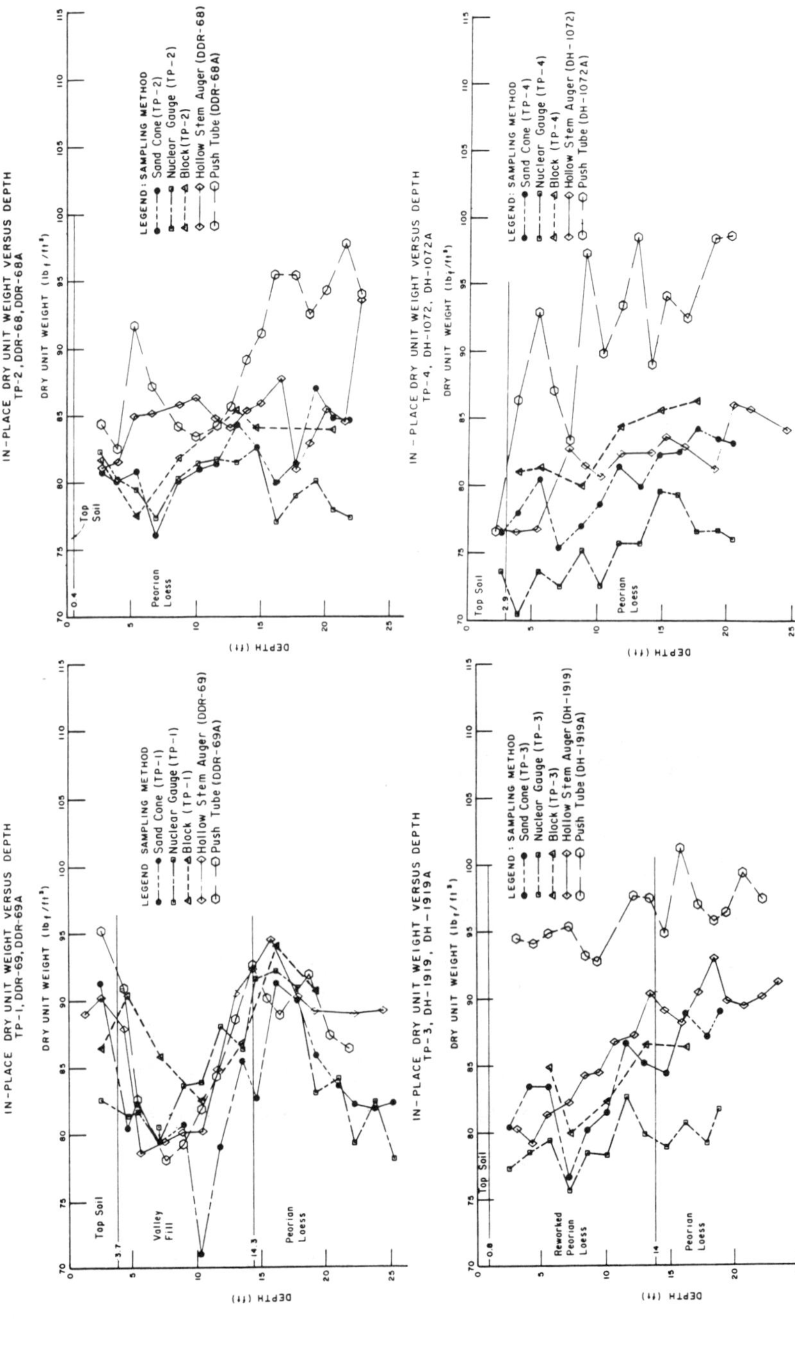

Figure 8. - In-place dry unit weight versus depth

indicated minimal disturbance (i.e., no concave fractures). Further investigations are required to develop an economical liner meeting necessary tolerances for sampling with the 6-1/4-inch (16-cm) i.d. hollow stem auger.

Push Tubes

Following completion of each hole drilled with the hollow-stem auger system, an adjacent hole (offset several feet) was drilled using 5-inch (13-cm) i.d. push tubes and a Failing 1500S drill rig.

Continuous sampling was performed throughout the depth of all four push-tube holes (DDR-68A and -69A and DH-1919A and -1072A), with push-tube samples being taken at approximately the same depth intervals sampled with the hollow-stem auger system. Moisture content and dry unit weight were determined on all samples obtained. Sample recovery was computed for all samples.

Sample disturbance was observed, and sample recovery was consistently low (see tables 1 through 4). Radiographic examination of several push-tube samples indicated that substantial disturbance occurred during the sampling process. Figure 9 shows a photograph of concave fracturing of a loessial soil sample obtained with the push tube. This type of fracturing is frequently encountered in loessial soil samples when push tubes are used for sampling. At the discretion of the driller, in an attempt to improve sample quality and recovery, the penetration rate of the push-tube sampler was greatly reduced during a portion of this investigation program. Even at slower penetration rates, sample disturbance was observed and sample recovery was low.

Figure 9. - Concave fracturing of loessial soil sample caused by push-tube sampling

The results of moisture content and dry unit weight determinations made on samples from all four push-tube holes are summarized in

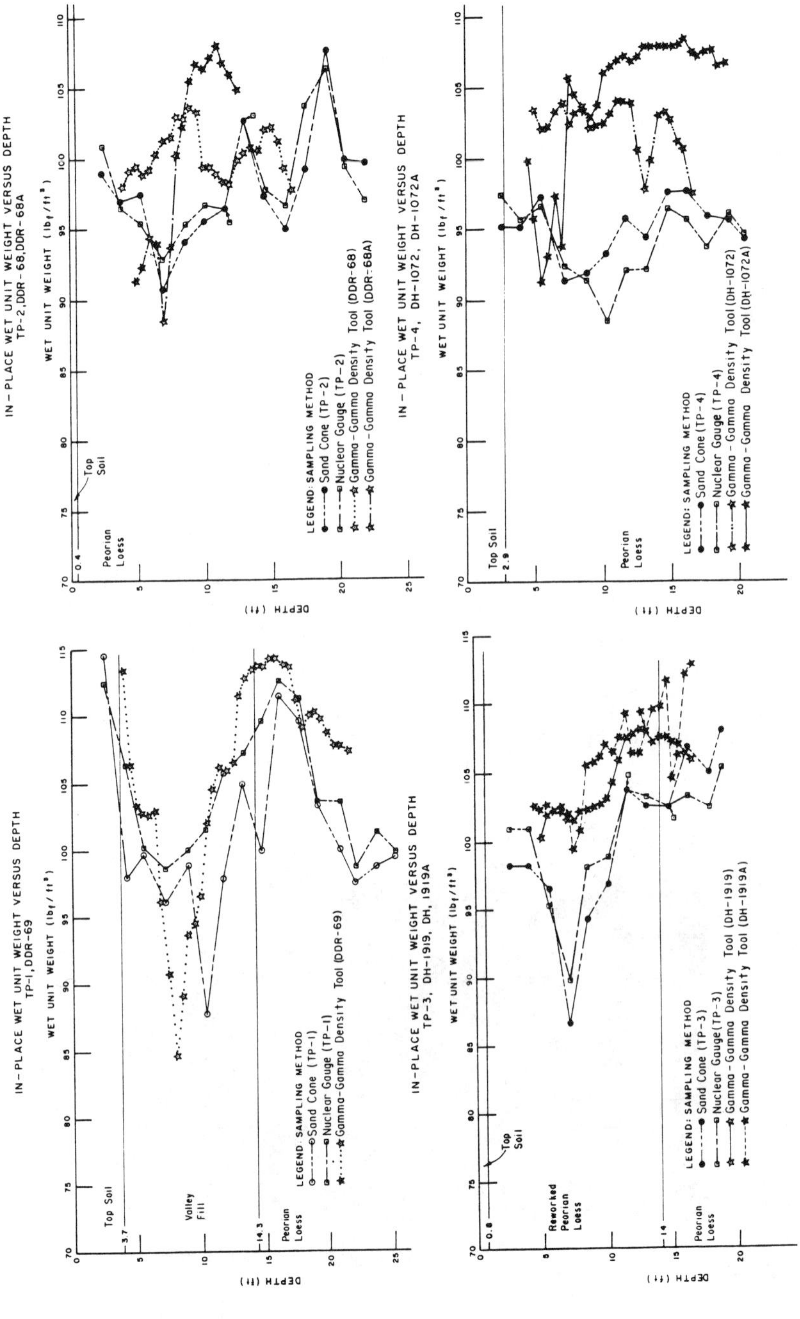

Figure 10. - In-place wet unit weight versus depth

tables 1 through 4. Plots showing dry unit weight versus depth are shown on figure 8.

Geophysical Borehole Unit Weight Logging

Following completion of sampling with the hollow-stem auger and push tubes, seven of the eight drill holes were geophysically logged. Drill hole DDR-69A caved in before geophysical logging could be performed. In-place wet unit weights were determined at frequent depth intervals. These results are shown on plots of in-place wet unit weight versus depth (fig. 10). Because only wet unit weight determinations can be obtained by geophysical borehole logging, these results are compared only to wet unit weights obtained by other sampling and testing methods.

In-Place Unit Weight Testing

Test Pits

Four test pits (TP-1 through -4) were excavated using an Insley H-600 backhoe. Total depths of the test pits ranged between approximately 19 and 25 feet (6 and 8 m). The test pits were excavated to full depth in an arc around the drill holes. Benches were excavated down one side slope of each test pit at approximately 1.5-foot (0.5-m) depth intervals for field unit weight (sand-cone) testing, surface nuclear moisture-density gauge testing, and block sampling. Figure 11 is a photograph of an excavated test pit. In-place moisture content and dry unit weight determinations were made on each bench by the sand-cone method and by nuclear gauge. Block samples were also obtained at frequent depth intervals in all test pits for laboratory moisture content and dry unit weight determinations by the mass in air - mass in water method.

Figure 11. - Typical test pit configuration

Surface Nuclear Gauge

Nuclear gauge moisture content and unit weight testing were performed on each bench in all four test pits at approximately 1.5-foot (0.5-m) depth intervals. A single probe nuclear gauge was used for testing. The gauge was adjusted once at each test pit to correct for sidewall proximity effects. Three 1-minute counts were taken at each location. The three readings were averaged to obtain the final reading. Results of nuclear gauge testing are summarized on plots shown on figure 8.

Because the moisture content is a factor in obtaining correct in-place dry unit weights, ovendried moisture content values were used. When possible, moisture content determinations from the sand-cone test were used; however, in several instances, separate moisture content determinations were required because of the length of time between nuclear gauge and sand-cone testing. Moisture contents of ovendried specimens and the wet unit weight values obtained by the nuclear gauge were then used to calculate in-place dry unit weights for the nuclear gauge. These results are shown on figure 12.

In-place wet unit weights obtained by the nuclear gauge were also compared to wet unit weights obtained by the sand-cone and gamma-gamma density tool testing. These results are plotted on figure 10.

Field Density (Sand-Cone) Tests

A sand-cone test was performed on each bench in all test pits following testing using the surface nuclear gauge. Results from the sand-cone tests are summarized on plots shown on figure 8. The sand-cone test was used as the standard, and data from all other test methods were compared to results from the sand-cone test.

Block Samples

Small block samples, having an approximate average volume of 0.14 ft^3 (4000 cm^3), were obtained from benches at frequent depth intervals in all test pits. The block samples were waxed at the site and were then transported to the laboratory for moisture content and dry unit weight determinations. The dry unit weight was determined by the mass in air - mass in water method. These test results are summarized on plots shown on figure 8.

RESULTS OF SAMPLING INVESTIGATION

In-Place Dry Unit Weight Versus Depth

Tables 1 through 4 and figure 8 summarize in-place dry unit weights determined from 5-inch (13-cm) i.d. push-tube samples, 6-1/4-inch (16-cm) i.d. hollow-stem auger samples, block samples, sand-cone testing, and surface nuclear gauge testing.

Davis Creek Damsite

In-place dry unit weights obtained from TP-1, DH-69, and DH-69A at Davis Creek damsite produced comparable data for most sampling intervals for all five methods (see fig. 8). The geologic log at this location indicated topsoil from the ground surface to a depth of 3.7 feet (1.1 m), Valley Fill between depths of 3.7 and 14.3 feet (1.1 and 4.4 m), and Peorian Loess between depths of 14.3 and 25.1 feet (4.4 and 7.7 m).

In-place dry unit weights obtained in Peorian Loess from TP-2, DH-68, and DH-68A, also at the Davis Creek damsite, show comparable values for four of the five methods (see fig. 8). The 5-inch (13-cm) push-tube samples generally showed higher in-place dry unit weights than samples obtained by the other methods. Several of the push-tube samples showed extensive disturbance when observed visually and based on radiographic examination. Concave fracturing was typically encountered (see fig. 9 for a photograph) in these samples. Sample recovery was low for many of the push-tube samples (see table 2), indicating that compaction may have caused the increase in unit weight.

Mirdan Canal

In-place dry unit weights obtained in Peorian Loess from TP-3, DH-1919, and DH-1919A (located along the Mirdan Canal alignment) show comparable values for sand-cone testing, hollow-stem auger samples, and block samples (see fig. 8). Again, samples obtained with 5-inch (13-cm) push tubes show consistently higher in-place dry unit weights than those obtained by the sand-cone method. Low recovery (see table 3) and extensive sample disturbance were frequently encountered with the push-tube samples, indicating compaction during the sampling process. At this location, in-place dry unit weights obtained by the surface nuclear gauge were lower than those obtained by the sand-cone method; however, wet unit weights obtained by the surface nuclear gauge were comparable to those obtained by the sand-cone method (fig. 10). This indicates that a field correction for moisture content is necessary when using the surface nuclear gauge.

In-place dry unit weights obtained from TP-4, DH-1072, and DH-1072A (also located in Peorian Loess along the Mirdan Canal alignment) show comparable values for sand-cone testing, hollow-stem auger samples, and block samples (see fig. 8). Again, samples obtained with 5-inch (13-cm) push tubes were disturbed and sample recovery was low(see table 4). Consistently higher in-place dry unit weights were obtained from 5-inch (13-cm) push-tube samples than from the other methods, indicating compaction during the sampling process. In-place dry unit weights obtained from the surface nuclear gauge were again consistently lower than those obtained by the sand-cone method; however, the wet unit weights were comparable to those obtained by the sand-cone method (fig. 10). This indicates that a field correction for moisture content is necessary when using the surface nuclear gauge.

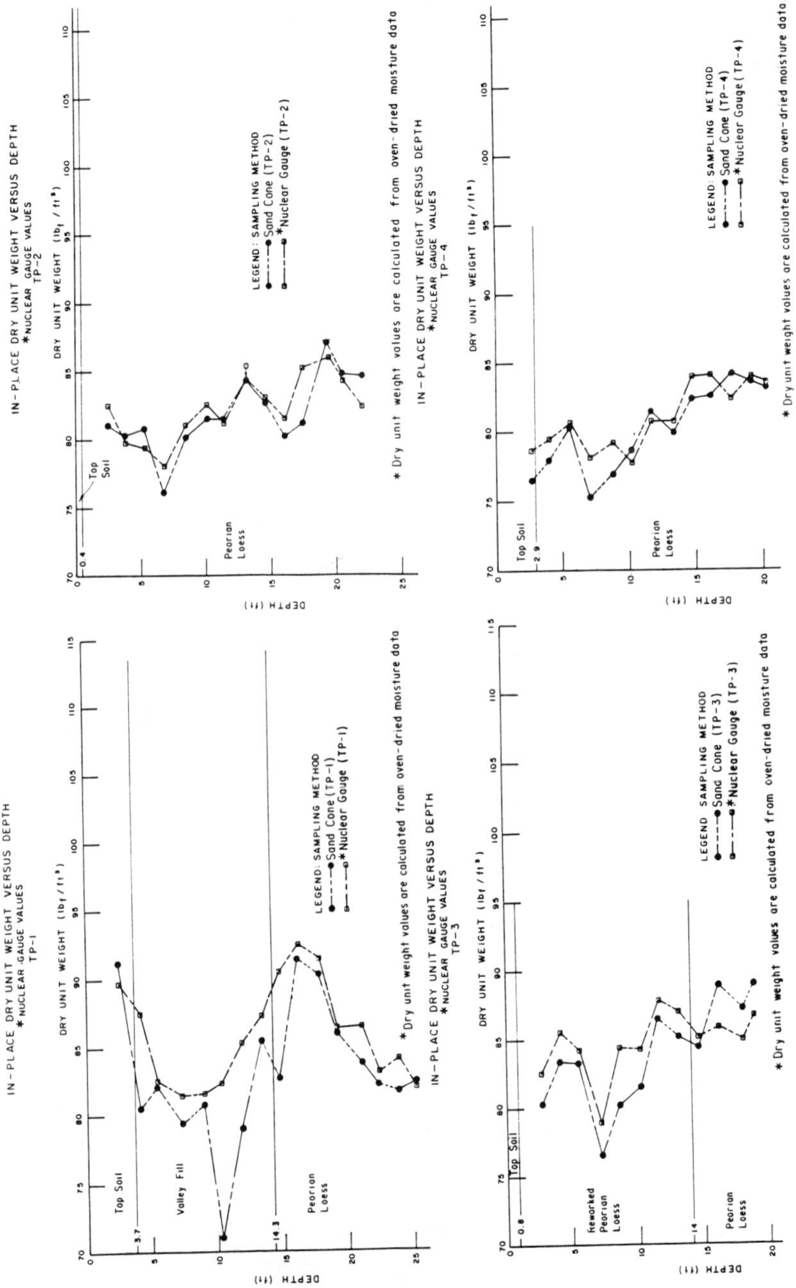

Figure 12. - In-place dry unit weight versus depth - nuclear gauge

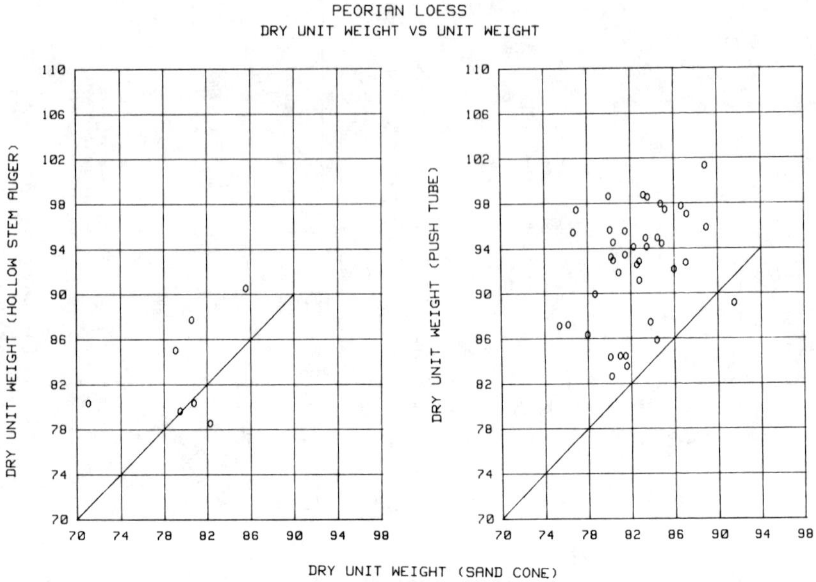

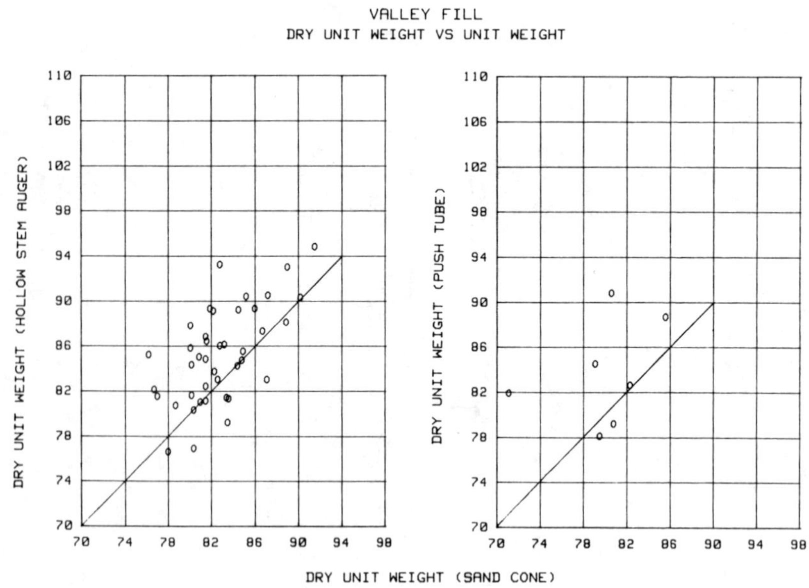

Figure 13. - Dry unit weight (hollow stem auger and push tube) versus dry unit weight (sand cone)

When comparing results of in-place dry unit weight versus depth at all four locations (fig. 8), results generally indicate that in-place dry unit weights obtained by the sand-cone method, the hollow-stem auger, and block samples produce comparable values. In-place dry unit weights obtained from 5-inch (13-cm) push-tube samples were consistently higher [up to 20 lb/ft^3 (3 kN/m^3)] than values obtained by the sand-cone method. Low recovery (tables 1 through 4) and extensive sample disturbance were frequently encountered in 5-inch (13-cm) push-tube samples, indicating compaction during the sampling process. Because the amount of compaction which occurred during the push-tube sampling process is unknown, the change in unit weight due to compaction cannot be determined. Therefore, dry unit weights obtained from push-tube samples are probably not representative of in situ soil conditions. In most cases, in-place dry unit weights determined from hollow-stem auger samples were reasonably close to values obtained by the sand-cone method. Sample recovery was high (tables 1 through 4), and quality was good when the hollow-stem auger was used for sampling.

Plots showing dry unit weights determined by the hollow-stem auger and push tubes versus dry unit weights obtained by the sand-cone test are shown on figure 13. Unit weights determined from samples of Valley Fill and Peorian Loess are plotted separately.

In-place dry unit weights obtained by the surface nuclear gauge were frequently lower than values obtained by the sand-cone method; however, wet unit weights obtained by the surface nuclear gauge were reasonably close to wet unit weight values obtained by the sand-cone method. This indicates that a field correction for moisture content is necessary when using the surface nuclear gauge. As part of this investigation program, moisture content samples were obtained for ovendrying at all locations where the surface nuclear gauge was used. Ovendried moisture contents and wet unit weights obtained from the nuclear gauge were used to compute in-place dry unit weights determined by the nuclear gauge and are summarized on plots shown on figure 12. These in-place dry unit weight values were generally comparable to those obtained by the sand-cone method.

PRACTICAL APPLICATIONS

The loose structure of loessial soil makes it difficult to obtain "undisturbed" samples for in-place unit weight determinations and laboratory testing. Sampling in the dry is necessary for loess, as drilling fluid can penetrate the sample, weakening the soil structure and consolidate the sample when pressure is applied during the drilling process. Good quality samples can be obtained from test pits; however, when samples are required at depths greater than practical for test pit excavation, other sampling techniques are required. The hollow stem auger system with liners provides a practical and economical method for obtaining good quality loess samples for unit weight determinations and laboratory testing.

CONCLUSIONS AND RECOMMENDATIONS

1. In most cases, in-place dry unit weights obtained from hollow-stem auger samples were reasonably close to values obtained by the sand-cone method. Sample recovery was consistently high, and sample quality was good when the hollow-stem auger was used for sampling.

2. In-place dry unit weights obtained from 5-inch (13-cm) push-tube samples were consistently higher [up to 20 lb/ft^3 (3 kN/m^3)] than values obtained by the sand-cone test. Low recovery and extensive sample disturbance were frequently encountered in the 5-inch (13-cm) push-tube samples. In addition, there appeared to be some sloughing of the drill hole side walls. For these reasons, there is no method for determining the amount of compaction which occurs during the push-tube sampling process, so the change in dry unit weight due to compaction cannot be determined. Therefore, dry unit weights obtained from push-tube samples are probably not representative of actual in-place soil conditions. These samples are generally very disturbed; and if other laboratory tests are performed on the material as sampled, the results may be inaccurate.

3. In-place dry unit weights obtained by the surface nuclear gauge were frequently lower than those obtained by the sand-cone method; however, wet unit weights obtained by the surface nuclear gauge were reasonably close to wet unit weight values obtained by the sand-cone method. This indicates that a field correction for moisture content is necessary when using the surface nuclear gauge to provide acceptable in-place dry unit weight data.

4. In most cases, in-place dry unit weight determinations made on block samples correlated closely with data obtained by the sand-cone method.

5. The gamma-gamma density tool generally provided higher wet unit weights than did the sand-cone and nuclear gauge. However, the gamma-gamma density tool frequently indicated the zones of low unit weight material determined by sand-cone and nuclear gauge testing.

6. Current USBR practice in loessial soil requires that soil samples be obtained either as block samples from test pits or with the 6-1/4-inch (15.6-cm) i.d. hollow-stem auger system to ensure that the actual in-place soil conditions are represented by the samples. Because of the loose structure of loessial soil, samples obtained by the hollow-stem auger for unit weight and laboratory testing should be limited to a maximum length of 1.5 feet (0.5 m) to minimize sample disturbance. Further studies may indicate that longer undisturbed samples can be obtained with the hollow-stem auger for laboratory testing.

7. Further studies need to be initiated to develop an economical liner meeting necessary tolerances for sampling with the 6-1/4-inch (15.6-cm) i.d. hollow-stem auger system. PVC pipe sections were used in this investigation program; however, there was substantial variation in the i.d. of the pipe making it difficult to obtain a proper fit with the sampler bit.

REFERENCES

Earth Manual, 2d edition, U.S. Department of the Interior, Bureau of Reclamation, U.S. Government Printing Office, Washington, D.C., reprint 1980.

Gibbs, H. J., and W. Y. Holland, Petrographic and Engineering Properties of Loess, Engineering Monograph No. 28, Bureau of Reclamation, Denver, Colorado, November 1960.

ACKNOWLEDGMENT

The author wishes to acknowledge Larry D. Cast, Project Geologist, Raymond Kehler and Norman Prince, Civil Engineering Technicians, Nebraska-Kansas Projects Office, Bureau of Reclamation, Grand Island, Nebraska; and Richard McKenny, Chief, North Loup Division Laboratory, Bureau of Reclamation, Ord, Nebraska; for their valuable input and assistance in data acquisition. Also to Richard A. Young, Technical Specialist, Geotechnical Branch, Division of Research and Laboratory Services, for his technical assistance in the review of this paper.

Dynamic Characteristics of Intact Loess

Wu Zhi-hui*, Xie Ding-yi*, Yu Xiong-fei*

Li Qi-yao** and Chen Xian-yao**

Abstract

In this paper the experimental results of dynamic stress-strain relation, dynamic shear modulus, dynamic strength and dynamic deformation characteristics of loess obtained in cyclic triaxial apparatus are presented. On this basis some conclusions of practical and theoretical importance are summarized.

Introduction

Almost 13,000,000km^2 of ground surface all over the world is covered with the loess deposits. There is the loess of about 640,000km^2 in China. It is characterized by its wide distribution, complete stratum and large thickness. Owing to that the loess in foundation of structures or in natural slopes often suffers dynamic loading besides the action of static stress, the researches of dynamic characteristics of loess have brought to the investigators great attention due to the developing of soil dynamics. It must be recognized, however, that the study in this aspect is still in the early stage of its development and the published papers are now very poor. This paper is to take aim at going further into the strength and deformation aspects of loess and at giving the primary research results.

It is widely known that the loess has a series of peculiar behaviour differing from the other soil types because of its specific condition of formation. The typical loess is one of the aeolian soils formed in the regions under the conditions of dry and semi-dry climate, abundant salts and under-consolidation. It is characterized with superiority of silt-graded particles, low degree of moisture, small density and granular cement-frame structure. Hence, once its structure is destroyed by the actions of static force, water or dynamic loading, a large deformation causing danger to the safety or normal use of structure laid on or in the loess deposit could always take place. The development of such a deformation or destruction appears with a complex law due to the variety of magnitude, order and duration of above mentioned actions, as well as the difference in initial moisture, density and structure of loess under concrete conditions. It is undoubtedly that both the variety of inner factors such as the moisture, density and loess structure and the variety of outer factors such as the actions of static loading, water and dynamic loading, particularly the initial moisture and the action of water, should be considered in research of dynamic characteri-

* Xie Ding-yi---Prof., the others---Lecturers
Shaanxi Institute of Mechanical Engineering, Xian, China
** Xian Institute of Metallurgy and Construction Engineering, Xian,China

stics of loess.

Multiple sets of test on undisturbed loess samples from Xian, Lowchuan and Lanchow (Table 1) in the dynamic triaxial apparatus both of pneumatic pattern developed by Research Institute on Geotechnical Engineering in Shaanxi Institute of Mechanical Engineering and of DSD-electro-magnetic pattern developed by the Northwest Electric Power Design Institute have been carried out in order to research the dynamic characteristics of intact loess. All the loess samples can be classified as loessy sandy clay of geological age in Q_2 and Q_3. The samples from Lanchow and Lowchuan loesses are in a lower density and the latter has a lower water content than the former. The samples of Xian loess No.2 have a higher density, a wider range of initial water content which is generally higher than plastic limit, and a degree of saturation between 50%

Table 1. principle physic-mechanic characteristic index of loess tested

Index		Lanchow Loess	Xian Loess No.1	Xian Loess No.2	Lowchuan Loess
Dry density	KN/M^3	11.8-13.7	12.3-13.8	13.1-17.2	12.7-13.1
Initial water content	%	16.3-28.7	19.1-23.0	14.2-31.0	9.7-12
Void ratio	e	0.96-1.30	0.96-1.20	0.58-1.07	1.00-1.20
Plastic limit	w_p	16.3-19	17-19		19.7
Liquid limit	w_l	24.0-32	30-32		30.7
Shear strength	C, kpa	16-26	20		30-80
	φ, degrees	16-18	22		32-34

and 100%. In the tests different values of consolidation stress ratio K_c(1.0-2.0) are used. After the sample is consolidated under given principle stresses, the dynamic stress is then applied in the axial direction of sample and the time history of dynamic stress, dynamic strain and dynamic pore water pressure are recorded under undrained test condition. Here, the dynamic stress is applied step by step in increasing amounts for the research of dynamic stress-strain relation and modulus; but it is applied once in a given magnitude until the failure or a stationary deformation are observed for the research of dynamic strength and deformation. In order to minimize the amount of test the dynamic stress application of two stages having a large difference in magnitude is also adopted in the test aimed to research only the dynamic deformation. In this manner it may be considered that the small deformation caused by the former small dynamic stress would not have remarkable influence on the large deformations caused by the latter large dynamic stress. Other suitable procedures of dynamic stress application are also used in the tests to research the effect of the order of water, static and dynamic loading actions on the dynamic characteristics of loess samples. The results of the tests are presented as follows.

Dynamic Stress-Strain Relation

The form of dynamic stress-strain relation obviously depends upon the initial water content of intact loess. The test results of Lowchuan loess with low natural water content and high artificially increased water content are shown in fig.1a and fig.1b respectively. It may be seen that the dynamic stress-strain relation has a linear pattern for the samples of low water content and a hyperbolic pattern for those of high water content. The failure appears in a tensile brittle pattern and compression-shear pattern respectively. So the failure of soil with low water content is limited by the tensile strength with a failure strain about $1x10^{-3}$ to $2x10^{-3}$ and the dynamic stress at failure increases as the consolidation stress ratio K_c increases, while the failure of soil with high water content is governed by the shear strength. In the samples of Lowchuanloess, particularly of Xian loess with high initial water content the dynamic stress-strain relations of hyperbolic pattern are also noted. The mathematical models of above-mentioned relations for the dry and wet pattern of loess may be written as:

$$\sigma_d = c\,\varepsilon_d \tag{1}$$

and

$$\sigma_d = \frac{\varepsilon_d}{a + b\,\varepsilon_d} \tag{2}$$

where a, b and c are the test parameters depending upon the soil characteristics and consolidated stress state. They can be determined from the intercept and slope of line $\frac{1}{E_d} \sim \varepsilon_d$ in figure 1.(b), and the slope of lines in fig. 1 (a).

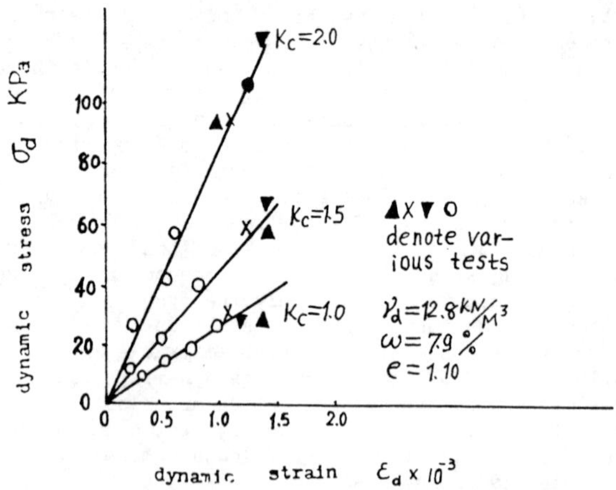

Fig.1.a. Relation between dynamic strain and dynamic stress of dry pattern loess

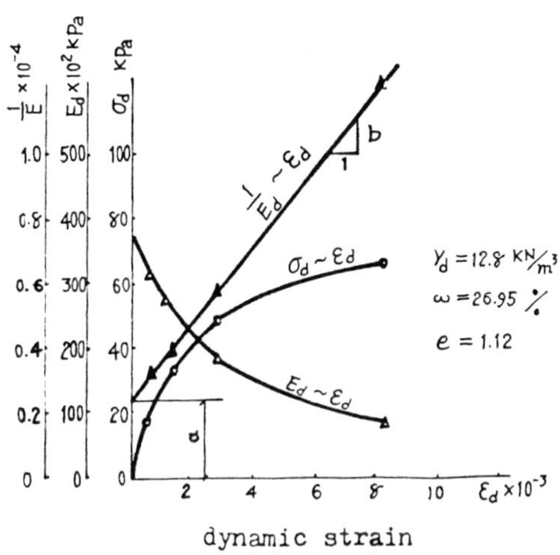

Fig.1 b Relation between dynamic stress and strain of wet pattern loess

Dynamic shear Modulus

According to the definition of modulus the following expressions may be easily obtained from the above-named models.

$$E_d = C = \sigma_d / \varepsilon_d \tag{3}$$

and

$$E_d = \frac{\varepsilon_d}{a + b\,\varepsilon_d} \tag{4}$$

The latter one can be rewritten as

$$E_d = \frac{1}{E_{d,max} + \dfrac{\varepsilon_d}{\sigma_{d,max}}} \tag{5}$$

or

$$\frac{E_d}{E_{d,max}} = \frac{1}{1 + \varepsilon_d \dfrac{E_{d,max}}{_{d,max}}} = \frac{1}{1 + \dfrac{\varepsilon_d}{\varepsilon_r}} \tag{6}$$

where $E_{d,max}$ is the maximum or initial modulus, $\sigma_{d,max}$ is the maximum axial dynamic stress, ε_d is the dynamic strain, σ_d is the dynamic stress and ε_r--the reference strain equal to $\sigma_{d,max}/E_{d,max}$.

The equations above may be written for the purpose of calculating the dynamic shear modulus by replacing ε_d, ε_r, $\varepsilon_{d,max}$ and E_d, $E_{d,max}$ with γ_d, γ_r, $\tau_{d,max}$ and G_d, $G_{d,max}$ respectively which are calaculated from the following relations:

$$\gamma_d = \varepsilon_d(1+\mu), \qquad \tau_d = \frac{1}{2}\sigma_d , \qquad \gamma_r = \frac{\tau_{d,max}}{\gamma_{d,max}} \tag{7}$$

where μ is poisson's ratio.

It is apparent that the dynamic modulus of loess decreases with the increase of dynamic strain amplitude and the degree of decrease depends upon the maximum or initial dynamic modulus and maximum or critical dynamic stress closely related to the soil characteristics and the initial stress state. The maximum dynamic stress is generally evaluated by the formula obtained on the basis of Mohr-Couloumb theory, but to evaluate the maximum dynamic modulus is more complex. The authors solve this problem by the following two ways: the one—to modify the well-known formula of Hardin and the other—to establish a new emperical formula.

Now, let's give some observation to Hardin's formula suggested for clay. Hardin's formula in the form[1]

$$G_{d,max} = 3230 \frac{(2.973-e)^2}{1+e} \sigma_m'^{0.5} (OCR)^K \tag{8}$$

may be written as

$$G_{d,max} = A_G \cdot \alpha \cdot \beta \tag{9}$$

Where $G_{d,max}$ is the initial dynamic shear modulus in KPa; α and β are the influence coeffecients of void ratio e and average effective principle stress σ_m' ; A_G is the influence coeffecient of other factors including the soil sort and the soil structure, in $(KPa)^{0.5}$

It has been found[2] that the following expression

$$\frac{G_{d,max,1}}{G_{d,max,2}} = \frac{\alpha_1}{\alpha_2} = \frac{\beta_1}{\beta_2} \tag{10}$$

would be satisfied for every two conditions of test on samples of Xian loess No.1 or Lanchow loess. It shows that the coefficients α and β for e and σ_m' in Hardin's formula are also applicable for these loesses tested. The value of coefficient A_G given by Hardin for normally consolidated clay equals 3230, for overconsolidated clay equals 3230 $(OCR)^K$ which is larger than 3230. So it is expected that the value of coefficient A_G for loess defined as under-consolidated soil would be lower than 3230 which expresses the influence of its structure characteristics. This has been identified by the test results of average value of A_G shown in Table 2.

It is considered that an apparent influence on the magnitude of soil structure strength can be effected by the original cement,moisture state and deviator stress, so the comparison tests are conducted with the samples of Xian loess No.1 in different initial water content and different consolidation stress ratio. The results obtained are shown in fig.2. It can be seen that the value of A_G decreases with the rise of degree of saturation and increases with the increase of stress ratio K_c.

Table 2. The value of coefficient A_G for tested soils.

Soil	Lanchow	Xian No.1	Xian No.2
number of samples	8	20	20
Average value of A_G $(KPa)^{0.5}$	1506	1340	1632

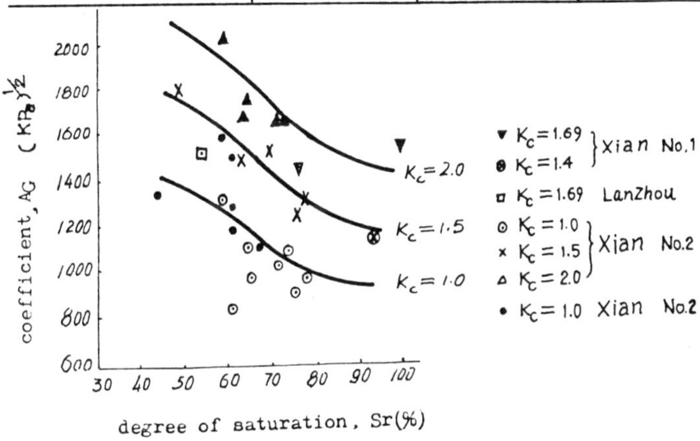

Fig.2(a) Relation between A_G and degree of saturation.

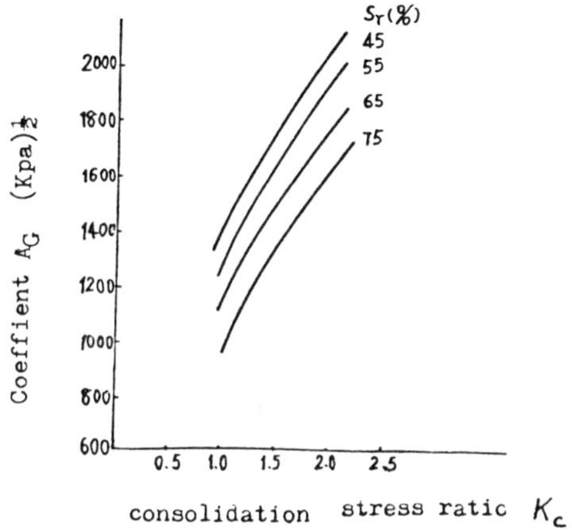

Fig.2(b) Relation between A_G and consolidation stress ration

The reducation of value A_G seems apparent when the degree of saturation is lower than 75%. That corresponds, to the water content ranging in the plastic state of soil. As in fig 2.b the magnitude A_G increasing with the stress ratio is less related to the change of degree of saturation and is proportional to $\sqrt{K_c}$. In fig2.a the test points of Lanchow loess and Xian loess No.2 are also plotted. It can be noted that in the coresponding conditions the value A_G of Lanchow loess is lower which denotes the influence of its lower structure strength, and the value A_G of Xian loess No.2 is higher which denotes the influence of its higher density.

The data in fig. 2a for Xian loess No.1 may be rearranged and normalized to obtain the curve in fig.3 which can be expressed by a parabolic as

$$\left(\frac{1}{\sqrt{K_c}}\frac{A_G}{A_{L1}}-1\right)^2=k(S_L-S_r) \tag{11}$$

Hence,

$$A_G = A_{L1}\sqrt{K_c}\left[1 + K(S_L-S_r)\right] \tag{12}$$

where

S_L-- degree of saturation cooresponding to liquid limit in %,

A_{L1}--coefficient A_G corresponding to liquid limit and K_c=1,

K -- test parameter.

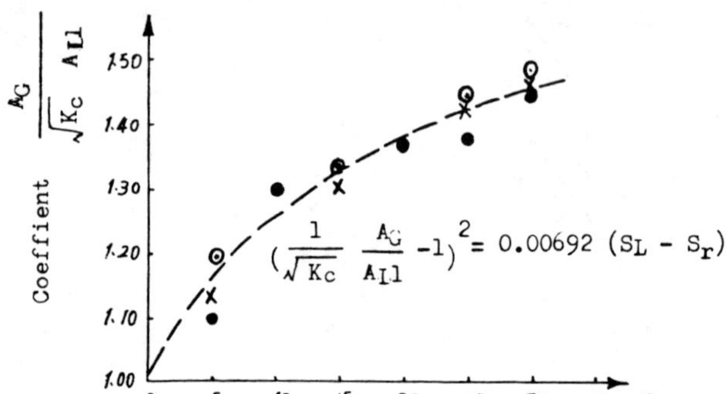

Fig.3 Relation between A_G/A_{L1} and (S_L-S_r)

It has been found for Xian loess No.1 that S_L=75%, A_{L1}=100(KPa)$^{0.5}$ and k=0.00692. Though these parameters for Xian loess No.2 have not been obtained because of its narrow distribution of degree of saturation near to the saturated state and of the stress ratio in tests only at K_c=1.69, the value of modulus $G_{d,max}$ calculated by formula (9) with the average coefficient A_G is close to that tested(Table.3). The ratio of tested value to the calculated value of $G_{d,max}$ for 18 samples equals 1.036 in average.

Table 3. Comparision of $G_{d,max}$ of Xian loess No.2 between tested and calculated values.

No of sample	water content w %	dry density d KN/M^3	Void ratio e	degree of sat. S_r %	effective mean stress σ'_m (KPa)	coeff. A_G (KPa)$^{\frac{1}{2}}$	Modulus $G_{d,max}$ KPax10^2 tested	formula (9) **	formula (13)
3-1	20.40	15.5	0.75	73.6	32	1498	239	260	176
3-2	14.20	13.1	1.07	51.6	72	3258	428	—	503
3-3	17.2	15.0	0.81	57.4	122	2340	667	465	677
3-4	28.7	14.5	0.87	89.3	174	1379	429	507	549
3-5	31.0	14.7	0.84	99.6	208	1736	618	580	638
3-6	23.6	16.2	0.67	95.1	226	912*	434	—	794
3-7	28.6	15.2	0.78	99.3	246	1728	729	688	770
3-8	27.5	15.9	0.70	100	267	1673	829	808	679
3-9	23.6	16.3	0.66	96.9	279	1775	953	876	945
3-10	23.1	16.4	0.65	96.3	294	1669	934	913	994
4-1	16.6	13.7	0.98	45.9	27	1740	181	169	172
4-2	23.4	15.1	0.80	79.4	72	1423	316	362	304
4-3	31.3	14.4	0.88	96.4	89	1155	253	357	324
4-4	30.2	14.4	0.88	92.9	102	1960	460	788	351
4-5	24.5	16.1	0.68	97.8	127	1154	406	574	495
4-6	21.2	17.2	0.58	98.9	150	1592	705	723	610
4-7	25.3	15.9	0.70	97.9	166	1576	613	645	603
4-8	23.1	16.2	0.67	93.4	215	1194	554	757	764
4-9	24.3	16.6	0.63	100	218	1640	813	808	790
4-10	22.6	17.0	0.60	100	247	2139	853	651	895
					Average value of coefficient A_G	1632			

* The max. and min. values are discarded in calculation of A_G

** calculated with the average value of A_G

The other way the authors adopted is to establish a new expression on the basis of experimental results about effect of every single factor, such as the void ratio e, the mean effective consolidation stress σ_m' and the degree of saturation S_r. It has been found for Xian loess No.2 that the initial dynamic modulus $G_{d,max}$ is directly proportional to (1.92-e), $(\sigma_m'/\sigma_0')^n$ and $e^{-0.1(S_r-60)}$ as shown in fig 4, fig 5 and fig.6. The final expression for determinating modulus $G_{d,max}$(in KPa) may be written as

$$G_{d,max}=32500\ Ds(1.92-e)\left(\frac{\sigma_m'}{\sigma_0'}\right)^n \qquad (13)$$

where

n—test constant, equal to 0.72-0.90 for k_c=1.69-2.0 and a smaller value for K_c=1.0~1.2. It can be adopted equal to 0.8 and 0.5 in calculations respectively.

k_s—influence coefficient of degree of saturation taking from Table 4.

σ_0'—normalize stress equal to 100 Kpa

Table 4. The value of coefficient K_s

S_r	50	55	60	65	70	75	80	85	90	95	100
K_s	2.631	1.989	1.600	1.364	1.221	1.134	1.080	1.049	1.030	1.018	1.011

The calculated values of $G_{d,max}$ from the formula (13) above for Xian loess No.2 are presented also in Table 3. A good agreement is obtained between the values $G_{d,max}$ tested and calculated.

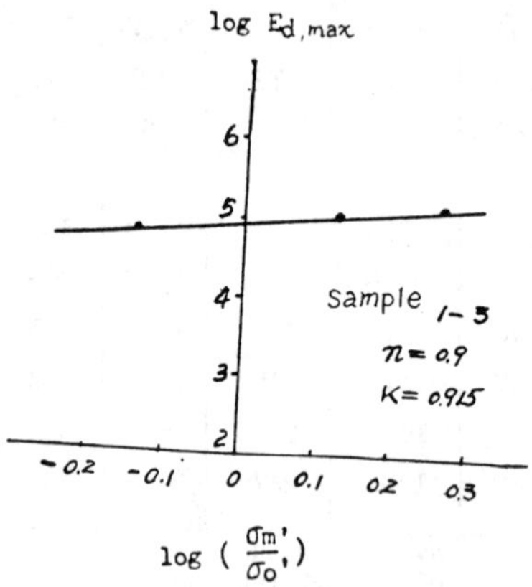

Fig.4 Relation between log E_0 and log $\left(\frac{\sigma_m'}{\sigma_0'}\right)$

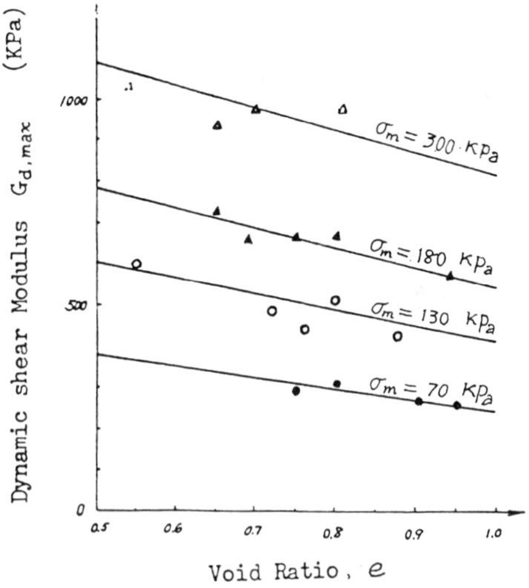

Fig.5 Relation between shear modulus $G_{d,max}$ and void ratio e

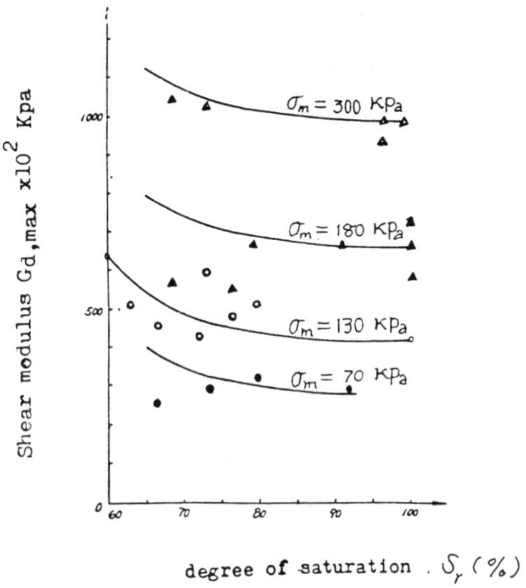

Fig.6 Relation between G and S_r shear modulus and degree of saturation

Dynamic Strength Characteristics

As a rule, dynamic strength is defined as the dynamic stress needed for producing a certain failure strain under the action of certain number of cycles. But application of this criterion for intact loess would meet a practical difficulty. For instance, the failure of Lowchuan loess of low water content is characterized by the brittle tensile crack with small dynamic strain and lack of residual strain and pore water pressure as shown in fig.7. This gives the reason for considering the dynamic strength of loess of dry pattern to be related to its tensile strength. The quickly developed soil crack will take place when the tensile strain increases gradually to a critical strain or the tensile strength decreases due to the action of dynamic stress to such a value as the applied dynamic stress. Because that the shear strength of soil is much higher than its tensile strength, the dynamic strength of loess of dry pattern is governed by its tensile strength and increases with the increase of consolidated stress ratio (fig.8), but does not distinctly decrease with the increase of number of cycles at failure N_f.

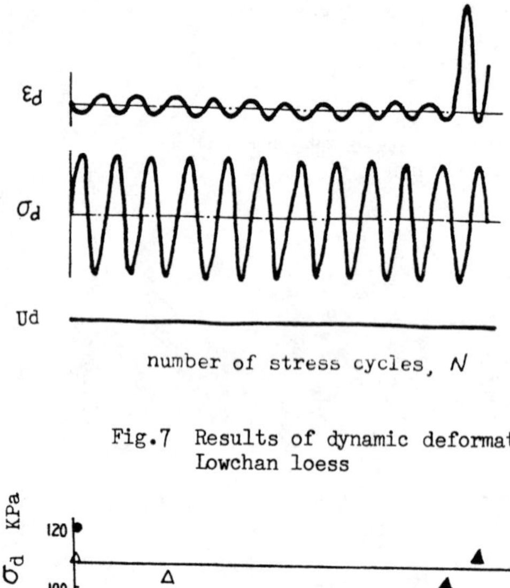

Fig.7 Results of dynamic deformation for Lowchan loess

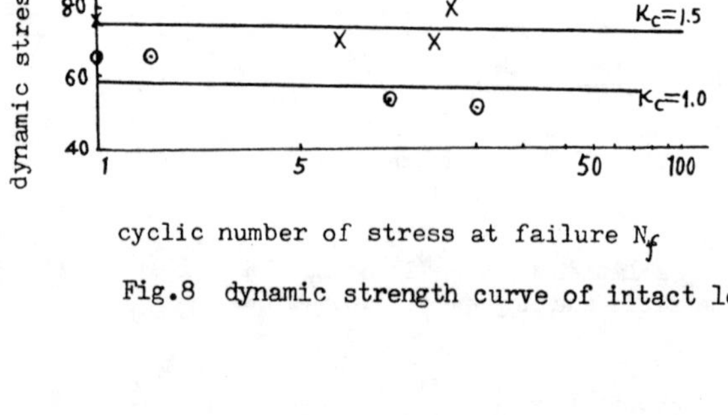

Fig.8 dynamic strength curve of intact loess

Let's take the saturated loess into account. The typical record of dynamic test is given in fig.9. It gives the similar behavior as that for the dense sand, i.e., the dynamic strain and pore water pressure increase with the increase of number of cycles. The time history of strain provides yield characteristics. The yield strain changes in wide range for different values of dynamic stress. This also makes it difficult to determine the dynamic strength with a given value of failure strain. At this time, it might be more reasonable to take the dynamic stress at the yield point with minimum curvature on ε_d-logN curves as the dynamic strength as shown in fig. 10.a. The dynamic strength determined with such a method for Lowchuan loess is plotted in fig. 11a. It shows the similar law as that determined with given failure strains 5% and 10%, but has a lower value than that determined with the given failure strains. It is necessary to note that the dynamic strength does not always increase but decreases with the increase of consolidation stress ratio K_c. This may be explained by the greater destruction of soil structure caused by the greater consolidation stress ratio. However it has also been observed that the magnitude of such a decrease in dynamic strength is reduced when the consolidation stress ratio continuously increases to a higher value because a new compression of loess is accompanied with the soil structure destruction.

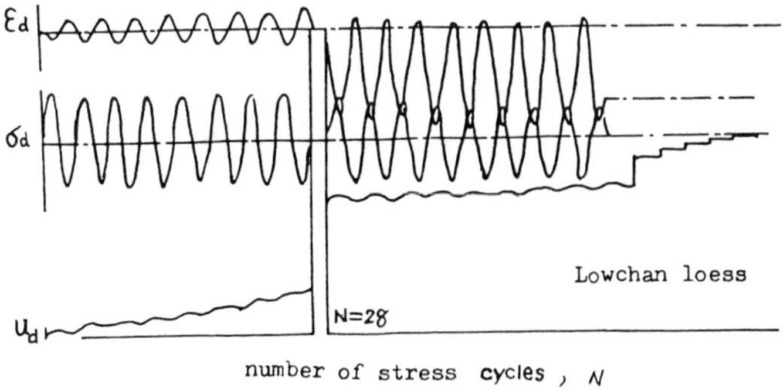

Fig.9 typical pattern of dynamic strength test (wet pattern)

If the time history of pore water pressure change is examined, then it may be seen from the $U_d \sim$logN curve in fig.10.b that for all the samples the pore water pressure does not reach to the value of lateral stress in spite of that the similar outward appearance to the liquefaction of saturated sand is observed. The pore water pressure level at the end of test rises continuously until it reaches a certain new level and then maintains at a stable value. This value is higher for isotropic consolidated condition than that for anisotropic consolidated condition, i.e., it reduces with the increase of the consolidated stress ratio. It is necessary to recognize that a certain stress is still undertaken by the soil skeleton and the pore water pressure is not uniformly distributed within the sample, so a transmission and transformation of the pore water pressure takes place after ceasing the dynamic action. Since the pore water pressure level at yield for different tests reaches only to a relatively low value, the failure of saturated loess may be considered to be caused principally by the increase of strain under dynamic loading.

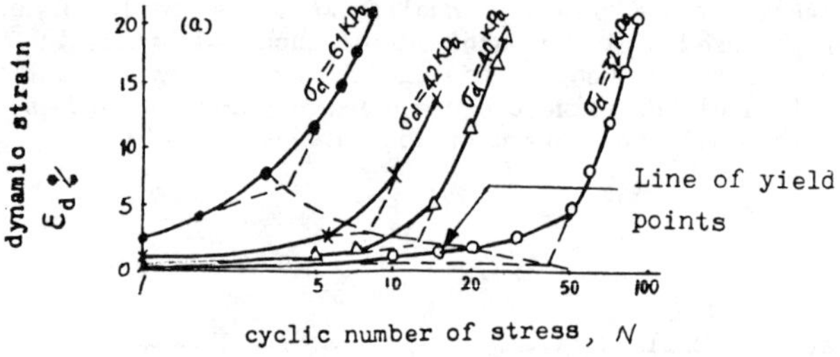

Fig.10 a) Relation between dynamic strain and number of stress cycles

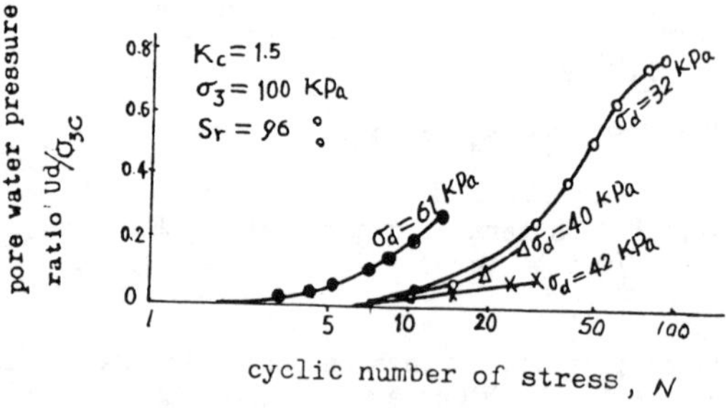

d) Relation between pore water pressure ratio and number of stress cycles

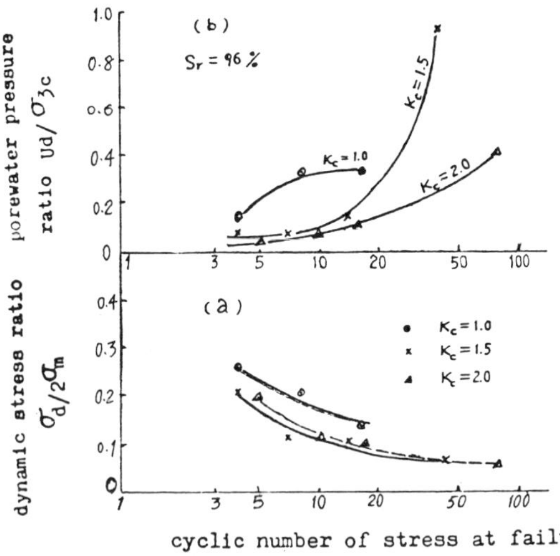

Fig.11 (a) dynamic strength curve of saturated loess
(b) pore water pressure ratio at failure

Dynamic Deformation Characteristics

The dynamic compression, as mentioned previously, must be considered as the center of dynamic deformation research of loess. It is related not only to the initial moisture state and the initial static stress state, to the magnitude and duration of dynamic stress, but also closely to the different action path of dynamic stress with respect to the actions of water and static loading. The dynamic deformation characteristics is generally illustrated by a plot of dynamic compression in terms of the dynamic stress as shown in fig. 12 and rearranged from it fig.13 for different number of cycles and different initial state which includes the moisture, density and stress state formed by the different action path before application of dynamic loading. It may be plotted from the measured residual deformation of several samples with the same initial state (γ_d, w and K_c) at a certain number of cycles. It shows that the dynamic deformation increases with the increase of dynamic stress and has a smaller amount of increase when dynamic stress varies in an extent of its low value, but has a larger amount of increase when the dynamic stress exceeds a certain critical value $\sigma_{d,cr}$. This critical dynamic stress depends upon the initial state of sample as shown in fig. 14. It may be seen that the critical dynamic stress reduces with increase of soil moisture. If this is investigated on account of the Atteberg limit of water content, then the critical dynamic stress rapidly increases when the water content is lower than the plastic limit and it reaches a sufficiently large value when the water content touches approximately the shrinkage limit. Therefore, the shrinkage limit may be defined as a critical water content, i.e., such a water content under which no dynamic compression could be developed by the dynamic stress of general magnitude. This shrinkage limit can be taken as a boundary to divide the loess into dry and wet patterns. The two regions above and below the curve in fig.14 are regarded as the regions in which the apparent residual deformation of loess can and can not occur for the loess at corresponding moisture and dynamic stress respectively.

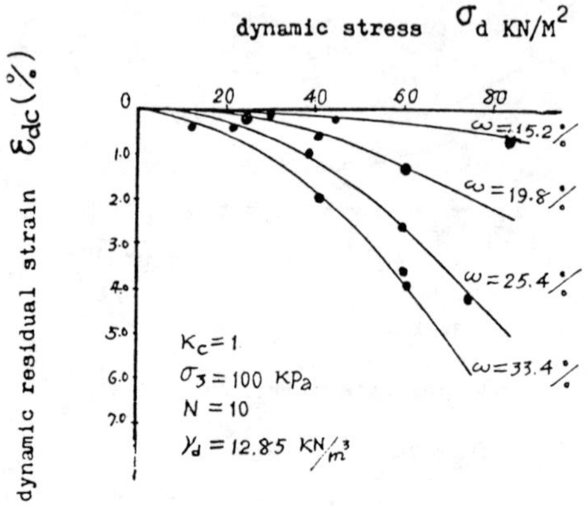

Fig.12 Relation between dynamic residual strain and dynamic stress

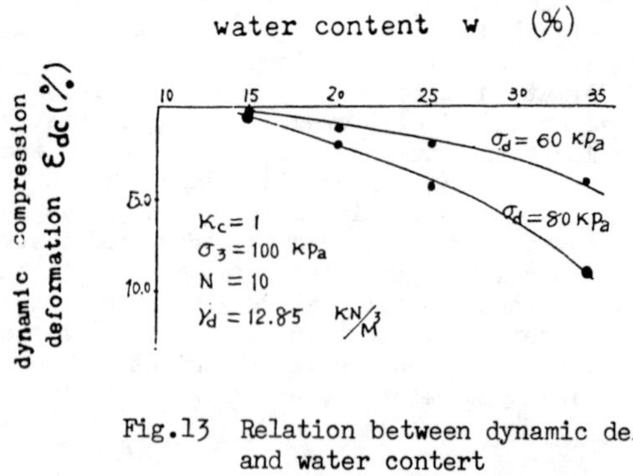

Fig.13 Relation between dynamic deformation and water contert

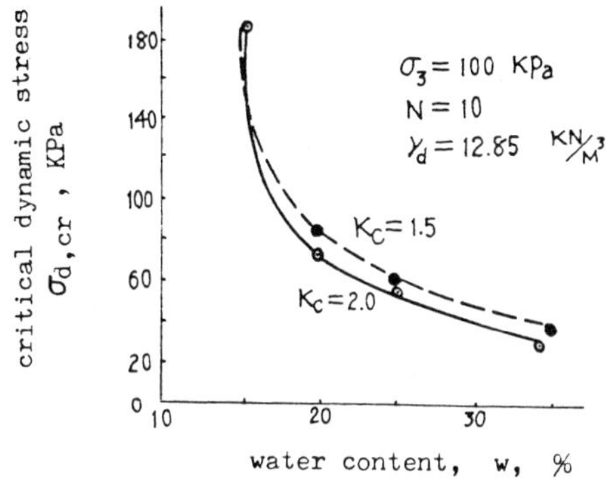

Fig.14 Relation between critical dynamic stress and water content

A series of tests has been carried out in the dynamic triaxial apparatus to research the effect of previous action of dynamic stress on the deformation of loess. In these tests the samples of loess are firstly consolidated under a stress ratio K_c equal to 2.0 and the static compressive deformations are obtained. Then samples are moistened to obtain moistening deformations which are plotted in fig.15 with white circle points as the fundamental of comparision. Consequently, the other unmoistened samples consolidated under the same condition are acted on with a certain dynamic stress to obtain the dynamic compressive deformations which are plotted in fig.15 with a wave-formed line. Then, the dynamic action on the samples is removed and the samples are moistened until their deformations become stationary to obtain the moistening deformations after previous dynamic action which are plotted in fig. 15 with triangular points. It may be seen from the results in fig.15 that the dynamic deformation due to the previous action of dynamic stress is generally equal to the decreased amount of moistening deformation occured after the action of dynamic stress. In another words, the previous action of dynamic stress replaces only a portion of action on the soil structure destruction that should have taken place originally due to the action of water. The dynamic compressive deformation due to such a portion of action of dynamic stress gradually increases with the increase of dynamic stress, while the other portion, i.e., the moistening deformation, being left to the action of water gradually decreases. The previous action of an appropriate dynamic stress may eliminate the moistening deformation for loess of higher water content when it is moistened after the action of dynamic stress. The general deformation, however, is still limited by the static moistening deformation. It is noteworthy that the deformation due to previous action of dynamic stress may much exceed the moistening deformation when the previous action of dynamic stress is sufficient enough to cause the strength failure of loess samples, and therefore, to cause the large lateral deformation of loess sample. The

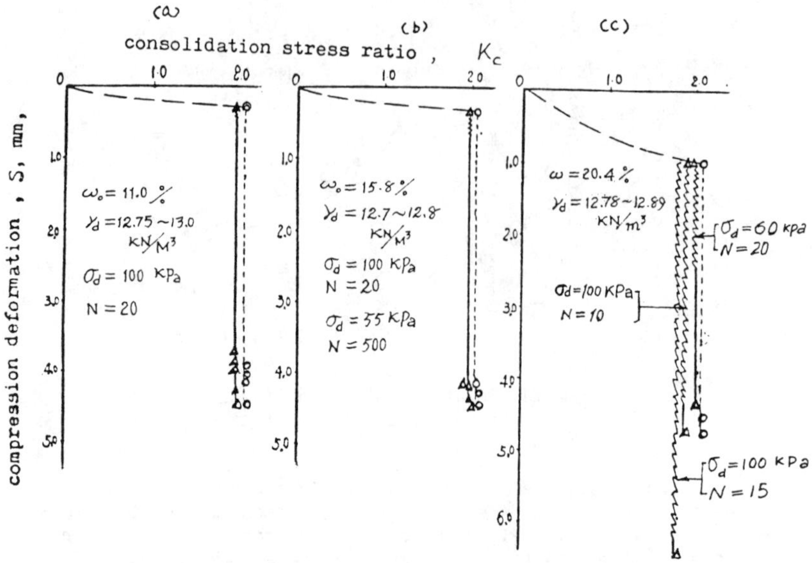

Fig.15 The influence of predynamic action on static moistening compression of Lowchan loess

------ static moistening compression without predynamic action

—— static moistening compression with predynamic action

~~~~ dynamic compression

— — — static compression

fundamental relations mentioned above can be summarized by the sketch in fig.16. Finally, the tests conducted under the condition of simultaneous moistening and dynamic stress indicate that the additional deformation defined as dynamic moistening collapse is related to the initial water content, the stress ratio, the mangnitude of dynamic stress and the number of cycles. In fig.17 the dynamic collapse strain determined by the bi-line method for Xian loess No.1 of different initial water content is presented in terms of dynamic stress. According to this method two samples are taken as a set, one of them is tested in saturated water content, and the other in natural water content. Both samples are suffering the same dynamic stress and the same number of cycles applied step by step. The difference between the deformations obtained in the two conditions is regarded as the dynamic moistening collapse. The plot in fig.17 shows that the higher the initial water content of loess sample, the lower the value of the dynamic moistening collapse, and the higher the consolidated stress ratio K_c, dynamic stress and number of cycles, the larger the dynamic moistening collapse. It is then necessary to indicate that a large dynamic moistening collapse might probably occur for the loess of low water content under the conditions of high stress ratio, large dynamic stress and high number of cycles. So it is a problem to be taken into account in practice.
~~~~

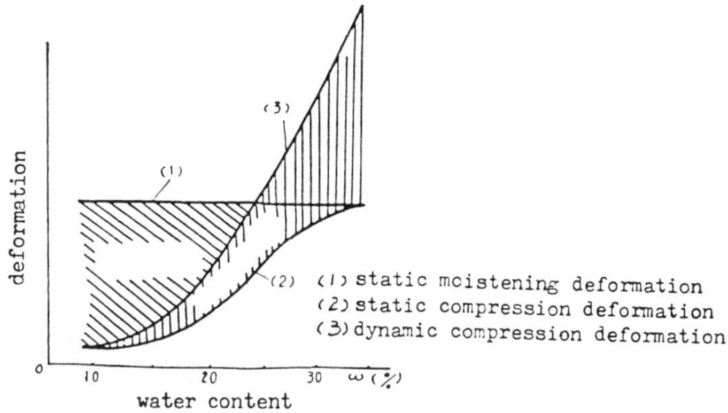

Fig.16 Relation between deformation and water content under different condition

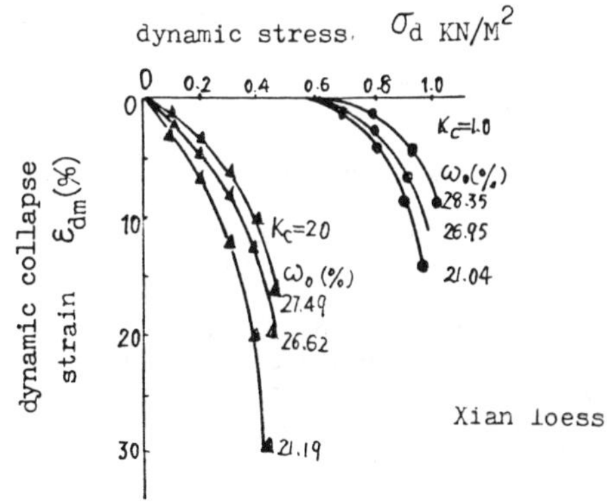

Fig.17 Relation between dynamic moistening collapse and dynamic stress

Conclusions

1. Special attention must be paid to the effect of loess structure and its destruction on the dynamic characteristics. The dynamic stress-strain relation, the dynamic strength and the dynamic deformation characteristics would be obviously changed and follow the different laws if the original loess structure is destroyed due to the action of high consolidation stress, moistening with water and dynamic loading. So it seems reasonable to divide the loess into dry pattern and wet pattern for their dynamic characteristic researches.

2. If the loess belongs to the dry pattern, i.e., its initial water content is lower than the shrinkage limit, then:

1) the dynamic stress-strain relation appears to be a linear pattern that has a constant dynamic modulus;

2) the dynamic strength is limited by the tensile strength of soil (brittle tensile crack) and its magnitude increases principally with the increase of consolidation stress ratio but independent of the number of cycles;

3) the dynamic compression is so small that it can be neglected in practice. But care must be taken with the dynamic crack and dynamic moistening collapse;

3. If the loess belongs to the wet pattern, i.e., its initial water content is higher than the shrinkage limit, then:

1) the dynamic stress-strain relation appears to be a hyperbolic pattern that has dynamic modulus decreasing with increasing magnitude of strain. The initial dynamic modulus for a given loess relates, in addition to the initial void ratio and the average consolidation stress, to the initial water content and stress ratio. It has a greater decrease when the initial water content changes within the plastic soil state;

2) the dynamic strength is limited by the shear strength (plastic compression-shear failure). The failure of loess occurs when the strain has accumulated to the yield strain that distinctly differs for different dynamic stress. The effect of consolidation stress ratio on dynamic strength depends on the degree of soil structure destruction caused by the consolidation stress. The dynamic strength decreases with the increase of consolidation stress ratio when the destruction of original soil structure due to the consolidation stress leads to a soil strength drop which exceeds the soil strength rise due to new compression. Otherwise, the dynamic strength increases with the increase of consolidation stress ratio K_c. The pore water pressure level at failure is lower and gradually rises to a new level when the dynamic stress has been removed. But it is still lower than the lateral stress and reduces with the increase of consolidation stress ratio.

3) the dynamic compression due to the action of dynamic stress is related not only to the magnitude of static and dynamic stress, but also to the history of action of water and to static and dynamic loading prior to the dynamic compression. This occurs because of the partial or entire destruction of loess structure during the mentioned actions. So long as such a destruction is not enough for the failure of soil strength, the different action path of above mentioned factors can only change the distribution proportion among static compressive, static moistening and dynamic compressive deformations with an invisible change of general deformation for the concrete condition.

4) with the increase of dynamic strain amplitude the dynamic modulus decreases cooresponding to the initial states of moisture, density and stress. The formulas suggested in this paper for evaluating the initial shear modulus permit consideration of these factors and calculating the required modulus of loess.

References

1. Hardin B.O., Drenevich, V.P. Shear Modulus and Damping in Soil, Journal of the Soil Mechanics and Foundation Division ASCE. Vol98 SM6, 1972 and Vol.98 SM8.1972
2. Xie Ding-yi, Wu Zhi-hui, Guo Yao-tong, Priminary Research of Intact Loess, China National Conference on Soil Dynamics, 1980.

SUBJECT INDEX

Page number refers to first page of paper.

AUTHOR INDEX

Page number refers to first page of paper.